SpringerBriefs in Applied Sciences and Technology

Series Editor

Andreas Öchsner, Griffith School of Engineering, Griffith University, Southport, QLD, Australia

SpringerBriefs present concise summaries of cutting-edge research and practical applications across a wide spectrum of fields. Featuring compact volumes of 50 to 125 pages, the series covers a range of content from professional to academic.

Typical publications can be:

- A timely report of state-of-the art methods
- An introduction to or a manual for the application of mathematical or computer techniques
- A bridge between new research results, as published in journal articles
- A snapshot of a hot or emerging topic
- An in-depth case study
- A presentation of core concepts that students must understand in order to make independent contributions

SpringerBriefs are characterized by fast, global electronic dissemination, standard publishing contracts, standardized manuscript preparation and formatting guidelines, and expedited production schedules.

On the one hand, **SpringerBriefs in Applied Sciences and Technology** are devoted to the publication of fundamentals and applications within the different classical engineering disciplines as well as in interdisciplinary fields that recently emerged between these areas. On the other hand, as the boundary separating fundamental research and applied technology is more and more dissolving, this series is particularly open to trans-disciplinary topics between fundamental science and engineering.

Indexed by EI-Compendex, SCOPUS and Springerlink.

More information about this series at http://www.springer.com/series/8884

Shruthi Kashyap · Vijay Rao ·
Ranga Rao Venkatesha Prasad · Toine Staring

Cook Over IP

Cordless Smart Kitchen Appliance Architectures and Protocols

 Springer

Shruthi Kashyap
Software Technology
Delft University of Technology
Delft, Zuid-Holland, The Netherlands

Vijay Rao
Software Technology
Delft University of Technology
Delft, Zuid-Holland, The Netherlands

Ranga Rao Venkatesha Prasad
Software Technology
Delft University of Technology
Delft, Zuid-Holland, The Netherlands

Toine Staring
Philips Research
Royal Philips NV
Eindhoven, The Netherlands

ISSN 2191-530X   ISSN 2191-5318   (electronic)
SpringerBriefs in Applied Sciences and Technology
ISBN 978-3-030-85835-3   ISBN 978-3-030-85836-0   (eBook)
https://doi.org/10.1007/978-3-030-85836-0

This Springer imprint is published by the registered company Springer Nature Switzerland AG
The registered company address is: Gewerbestrasse 11, 6330 Cham, Switzerland

Contents

Chapter 1
Introduction

Every person needs to eat regardless of whether he/she follows a "live to eat" or a "eat to live" philosophy. As food intake is an inevitable aspect of survival, food and everything around food have continuously evolved from time immemorial. One of the major aspects that has been experiencing much technological innovation is cooking, and in a broader sense, the kitchen. With lives of people becoming busier every day, the kitchen will certainly be the focal point of innovation in the near future.

With the emergence of the Internet of Things (IoT) technologies, the concept of 'Smart Kitchen' or 'Connected Kitchen' [1] is being developed. This concept has brought a wave of smart and connected devices that has transformed the way we cook and interact with our kitchen appliances. It facilitates many interesting and important applications that cater to the busy lifestyles of today, such as enabling the appliances to be controlled from smartphones, and cooking by uploading recipes from a remote location which saves time as compared to the conventional cooking methods.

An imminent technological development in the smart kitchen domain is the concept of 'Cordless Kitchen' [2]. This concept, introduced by the Wireless Power Consortium (WPC) [3], does not require the appliances to have power cords or batteries to operate. Instead, they are powered by inductive power sources (or power transmitters) that may be built into a kitchen counter, cooktop (hob), or a table. The appliance needs to be simply placed on top of the power transmitters and the user should be able to cook, interact and control the appliance remotely.

The Cordless Kitchen standard, also known as 'Ki', is based on the principles of 'Qi' wireless charging technology which is already prevalent in the market for charging smartphones. Ki, however, is designed for powering higher input wattage equipment. There are about 580 consortium members (as of March 2020) in WPC with many renowned Original Equipment Manufacturers (OEMs) such as Philips, Samsung and Robert Bosch. Some manufacturers are also extending this to charge laptops (e.g. Powermat Technologies).

As Ki does not deal with networking the appliances, this book focuses on getting these cordless kitchen appliances connected to the Internet with minimal changes on

S. Kashyap et al., *Cook Over IP*,
SpringerBriefs in Applied Sciences and Technology,
https://doi.org/10.1007/978-3-030-85836-0_1

the appliances (or devices) and networking stacks. In this chapter, we shall introduce the concept of cordless kitchens and technologies involved, and then give an overview of why connecting the appliances to the Internet is non-trivial.

The reader can learn about the 'Ki' cordless kitchen operation in detail in Chap. 2. We shall walk the reader through the complete design process: Chap. 3 shall discuss possible architectures to connect the appliances to the Internet; Chap. 4 shall describe the state of the art and present why this problem needs novel solutions; Chap. 5 details the challenges for providing Internet connectivity and describes how the TCP/IP protocol should be adapted to the cordless kitchen system. A thorough evaluation of the proposed solutions is presented in Chap. 6, along with few implementation recommendations. Other factors affecting the performance that a solution architect must consider are explained in Chap. 7.

1.1 Overview of the Cordless Kitchen Concept

The main goal of the cordless kitchen concept is to eliminate power cords in kitchen appliances. Connecting the appliances to the Internet would bring in ease of use. For example, users can upload recipes, monitor the dish and appliance and begin cooking when still on the way home. These goals together will provide the user with a truly wireless, smart-cooking experience.

1.1.1 Benefits of Cordless Kitchen

Ki is designed for cordless kitchen appliances that can be powered with a maximum of 2.2 kW. For heavy appliances, such as a refrigerator, that are stationary, the concept of a cordless kitchen is neither required nor efficient. Some of the benefits of the cordless kitchen are listed below.

- **Space efficient:**

  - Better usage of limited kitchen counter spaces and tables as the same space can cater to food preparation, cooking and cleaning.
  - Cordless appliances are easy to store.

- **Smart:**

  - Two-way communication between the appliance and the power transmitter allows for intelligent features such as consistent and power-efficient cooking, as the amount of power transferred is equal to what the appliance expects.
  - Adding Internet connectivity in the cordless kitchen would enable remote cooking, where users can control the appliances remotely, upload recipes and software updates, enable IoT communications, etc.

– Cookware can be made smart (e.g. smart pans).

- **Safe and Robust:**

  – Inductive power transfer is robust against water and dirt, as they have no effect on the operation or safety.
  – There is no electrical shock hazard.
  – The bottom of the appliances remains cool.
  – Foreign object detection functionality avoids accidental heating of foreign objects like spoons, knives, etc.
  – Leaving the power cord plugged in presents a safety hazard to some appliances. This can be solved by using the smart cordless kitchen.
  – If the appliance is knocked over or moved off the transmitter location, it immediately stops receiving power.

- **Interoperability:**

  – One standard power transmitter can be used for all types of cordless appliances.

- **Convenient and Clean:**

  – Reduces clutter in the kitchen.
  – As the power transmitters are installed underneath kitchen counters and tables, all the wiring is hidden from the view. This enables a sleek, ultra-modern design for the kitchen.
  – Kitchen countertop and the appliances are easy to clean.

- **Appliances can be moved:**

  – Enables table-top cooking.
  – Heat/cook where needed, efficiently. For example, appliances can be moved from the countertop to the dining table for keeping warm.

1.1.2   Use-Cases

Any kitchen appliance can be made cordless. A few examples include toasters, coffee makers, rice and slow cookers, deep fryers, etc. Three main use-cases are proposed by WPC [4].

Hybrid cooktops: In addition to the traditional hobs on an induction cooktop, one or two hobs will be Ki-compliant power transmitters. This enables versatile cooking along with traditional cookware when cooking multiple dishes. This is completely safe as the Ki-compliant cookware have their own integrated controls to prevent any undesired heating while the traditional cookware cannot draw any power from the power transmitter.

Kitchen counters: Installing a power transmitter under a kitchen counter enables a cordless kitchen. A variety of appliances can be placed on it allowing one to cook an entire meal on the countertop.

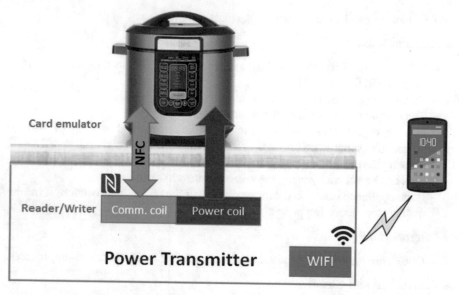

Fig. 1.1 The cordless kitchen concept

Dining tables: Making a dining table Ki-compliant allows for keeping the food hot and also cooking without any safety hazards (e.g. Chinese hotpots and cheese fondue) even with small children around.

1.1.3   System Architecture

In order to eliminate power cords, the appliance will be powered by inductive power transfer in which a permanently mounted Power Transmitter (PTx) or a Magnetic Power Source (MPS)[1] as shown in Fig. 1.1. The PTx contains a coil that draws power from the mains and transfers it via electromagnetic induction to another coil placed in the appliance [5]. The power is then converted back into electrical energy and/or heat for cooking within the appliance.

Interoperability: A Wireless Power Consortium (WPC) cordless kitchen-compliant appliance and power transmitter support interoperability, where one standard power transmitter can be used for all types of appliances. It is safe and robust as there will be no electrical shock hazard with inductive power transfer.

Safety: The Foreign Object Detection (FOD) functionality avoids accidental heating of foreign objects like spoons, knives, etc.

---

[1] PTx and MPS both represent the inductive power source and the terms are used interchangeably in the book.

Unlike traditional kitchen appliances, cordless kitchen appliances are made intelligent. They communicate with the PTx to ensure that the amount of power received remains within the limits of the appliance and according to the input from the user. The communication between the cordless appliance and the PTx takes place using a Near-Field Communication (NFC) [6] channel, as shown in Fig. 1.1. This makes cooking much more precise, responsive and repeatable with cordless appliances. The next section gives an overview of the NFC communication interface.

1.1.4 How Does it Work?

When an appliance is placed on the PTx, the PTx 'talks' to the appliance over the NFC channel in order to negotiate the amount of power to transfer. Here, the PTx operates in the NFC Reader/Writer (RW) mode and the appliance operates in the NFC Card Emulator (CE) mode, as depicted in Fig. 1.1. The communication is initiated and controlled by the PTx, i.e. it behaves as the master and the appliance as the slave. However, when it comes to power transfer, the appliance controls the amount of power it receives from the PTx by sending frequent power control messages.

The NFC technology, like the inductive power transfer, is also based on the concept of electromagnetic induction which enables short-range communication between two compatible devices. NFC operates with low magnetic field strength, and the presence of a high magnetic field corrupts the communication carrier of NFC. In the case of a cordless kitchen, both wireless power transfer and NFC communication need to work together in the same system. As the inductive power source generates a very high magnetic field that can disrupt the NFC communication, WPC has proposed a solution where the wireless power transfer and the NFC communication operate in a time-multiplexed fashion as shown in Fig. 1.2. Here, u_{op} represents the operating voltage and f_{op} represents the operating frequency. The NFC communication takes place at zero crossings of the power signal, for a duration of $T_{zero} = 1.5$ ms. For example, when a power signal with an operating frequency (f_{op}) of 50 Hz is used,

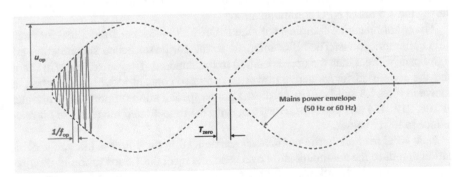

Fig. 1.2 Time-multiplexing of the power and the communication signals

Fig. 1.3 NFC antenna and primary coil in the PTx

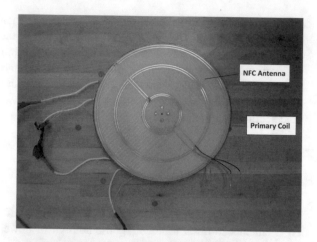

Fig. 1.4 NFC antenna and secondary coil in the appliance

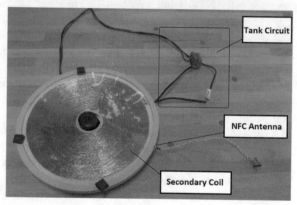

the power transfer would take place for 8.5 ms and the communication for 1.5 ms. This would repeat every 10 ms or in every half cycle of the power signal. It takes less than 1 μs to stop the power signal near the zero crossings and start the NFC communication. This avoids interference from the magnetic fields of the power signal during the 1.5 ms of NFC communication.

The communication complies with the ISO/IEC 14443 standard defined for near-field communications. The NFC antennas in the cordless kitchen specification are significantly larger than the ones specified in the standard. They are circular in shape and the coils used for inductive power transfer are concentric with each other, as shown in Figs. 1.3 and 1.4. The cordless kitchen specification supports NFC bit rates of 106, 212, 424 and 848 kbps. However, with the time-slotted mode, these bit rates reduce to lower values.

New NFC read and write commands are defined in the cordless kitchen specification to reduce the communication overhead and meet the 1.5 ms time slot requirement. The commands follow the ISO/IEC 14443 half-duplex transmission protocol. Table 1.1 shows the number of bytes that can be read using different NFC bit rates

Table 1.1 Number of bytes that can be sent using the new NFC read command at different data rates

Bit rate (kbps)	N° of bytes in payload
106	5
212	19
424	48
848	104

Table 1.2 Number of bytes that can be sent using the new NFC write command at different data rates

Bit rate (kbps)	N° of bytes in payload
106	4
212	18
424	47
848	103

in the time-slotted mode. The new write command supports similar payload sizes as shown in Table 1.2. These commands carry messages containing measurement data, operating limits, control data and auxiliary data for Internet connectivity. Further details about the NFC protocol extensions are provided in Chap. 2.

1.1.5 Internet Connectivity in the Cordless Kitchen

One of the requirements of the cordless kitchen is to enable Internet connectivity for users to control the appliances remotely, upload recipes and software updates, etc. Typically, the appliances need not support connectivity when they are away from the PTx. A straightforward way of providing Internet connectivity to the kitchen appliances would be to install Wi-Fi modules in them. However, in a cordless kitchen system, the appliance will not always be powered. Furthermore, when the PTx goes into standby mode, the appliance will not receive any power, i.e. it will be switched off. Therefore, the Wi-Fi module in the appliance will not be awake at all times to provide Internet connectivity, and would be available only when the appliance is placed on top of the PTx. This would lead to the loss of messages. Therefore, having a Wi-Fi module in every single kitchen appliance would be inefficient and unnecessary.

This research focuses on providing efficient Internet connectivity and enabling reliable communication with the appliances. We conclude this chapter with the challenges and proposed solutions in the next section.

1.2   Challenges and Solutions

Two main challenges are addressed through this research.

1 **Providing efficient Internet connectivity to the cordless kitchen appliances**
   On contrary to using a Wi-Fi module on each appliance, a Wi-Fi module or
   an Ethernet connection could be installed in the PTx, and the already existing
   NFC communication channel could be used to indirectly connect the cordless
   appliance to the Internet via the PTx. This would also make the appliances cost-
   effective as there can be only one Wi-Fi/Ethernet connection in the kitchen which
   could be used by all the appliances. Furthermore, when a message arrives onto
   the PTx, it can power up the appliance, if placed on the top of the PTx, and
   establish the communication.

   Two architectures are proposed in this book to provide Internet connectivity in the
   cordless kitchen through the NFC channel. Both these architectures are suited for
   the TCP/IP protocol stack. In the first architecture called the Proxy architecture,
   the appliance only sends the application data to the PTx via the NFC channel,
   and the PTx takes the responsibility of creating/processing TCP/IP packets and
   sending them to the end-user device. In the second architecture called the Bridge
   architecture, the appliance sends complete TCP/IP packets to the PTx, and the
   PTx only forwards these packets to the end-user device via the Wi-Fi channel.
   The book mainly focuses on the implementation and performance analysis of the
   bridge architecture where TCP/IP is tunneled over the time-slotted NFC channel.

2 **Adapting the TCP/IP protocol to a low bandwidth time-multiplexed NFC
   channel such that low end-to-end latency is maintained in the TCP applications**
   Most applications use TCP as the transport layer protocol in order to provide
   reliable end-to-end communications, and therefore it will be used in the appliances as well. However, NFC is designed for exchanging small payloads with
   data rates up to 848 kbps in normal mode and around 83.2 kbps in the time-
   slotted mode. On the other hand, TCP/IP is designed for exchanging a large
   amount of data and at much higher data rates to get a considerable performance.
   Tunneling a heavy-weight protocol like TCP/IP over a constrained channel like
   the time-slotted NFC would increase the system latency due to the large overhead introduced by the TCP/IP protocol with the TCP handshake/termination
   sequences, acknowledgment mechanism, header overheads, etc.

   The Internet applications of the cordless kitchen such as remote user control
   and online cooking are firm and soft real time as they require fast response
   time. Missing deadlines in these applications may not be hazardous, but it would
   definitely affect the cooking procedure and the quality of the food. Motivated by
   this demand, the research aims at adapting the TCP/IP protocol to the time-slotted
   NFC channel such that the TCP applications have low end-to-end latency.

   In this work, the feasibility of using the bridge architecture for firm and soft real-
   time applications is analyzed by studying performance bottlenecks and highlight-
   ing various factors affecting the latency, throughput and bandwidth utilization of

the NFC channel. Since the channel has unique properties, the challenges posed need to be solved. Several solutions and adaptation of TCP are proposed for the bridge architecture, as illustrated below.

a. Due to the delay on the NFC channel, TCP will experience spurious retransmissions.

To eliminate these, a generalized solution is provided using which appropriate TCP Retransmission Timeout (RTO) values can be calculated depending on the packet size and the data rate of the NFC channel being used.

b. Although TCP is designed to adapt the RTO over time by estimating the delay on the channel, it does not consider the payload sizes in this estimation. This leads to choosing an incorrect RTO value for this NFC channel. Therefore, if the payload size varies, TCP still experiences spurious retransmissions.

This research also proposes a new algorithm for dynamic RTO estimation for the TCP/IP packets considering the channel delays. This algorithm ensures that optimum RTO values are set for each packet such that spurious retransmissions are eliminated, and delayed retransmissions are prevented in case of packet loss.

c. The bridge architecture suffers from packet drops at the NFC interface due to the processing speed mismatch between the TCP/IP stack and the NFC module.

An NFC channel sensing mechanism is defined so that the TCP stack is slowed down to match the transmission speed of the NFC channel, thereby achieving an optimum inter-packet delay.

d. The other parameters of TCP, such as contention window size and maximum segment size, have an influence, which need to be studied. Also, the influence of bit errors in the NFC channel needs to be studied.

The book also throws some light on the parametric analysis of other factors that affect the system performance such as NFC bit error rates, communication time-slot sizes, presence of non-TCP/IP messages on the NFC channel, etc.

Using the new RTO estimation algorithm and the NFC channel sensing mechanism, a reduction of about 38% in the system latency is achieved at an NFC data rate of 11.2 kbps, and up to 53% at 24 kbps in the time-slotted mode. The methods to achieve this will unfold in the next chapters of this book.

1.3 Takeaways

The important takeaway message from this book is how to enable Internet connectivity to the cordless kitchen appliances despite the slow and time-slotted NFC channel. The minimal changes needed to the Internet stack are presented. Another takeaway message for the implementers/device manufacturers is what to expect when the networking parameters are tuned.

For the WPC consortium members and the standards working group, several possible architectures to connect appliances to the Internet are presented with pros and cons. We argue that a method to connect to the Internet should be included in the standards for better interoperability between the power transmitters and the appliances.

References

1. Smart Kitchen Gadgets | 2019 Guide to the Best Cooking Devices (2019). Postscapes. https://www.postscapes.com/connected-kitchen-products/
2. KI Cordless Kitchen Standard (n.d.). Wireless Power Consortium. https://www.wirelesspowerconsortium.com/kitchen/
3. Wikipedia contributors (2021). Wireless Power Consortium. Wikipedia. https://en.wikipedia.org/wiki/Wireless_Power_Consortium
4. Ki Cordless Kitchen: From Concept to Industry Standard (n.d.). Wireless Power Consortium. https://www.wirelesspowerconsortium.com/data/downloadables/2/3/7/5/ki-cordless-kitchen-white-paper-september-2019.pdf
5. Wikipedia contributors (2021). Electromagnetic induction. Wikipedia. https://en.wikipedia.org/wiki/Electromagnetic_induction
6. Wikipedia contributors (2021). Near-field communication. Wikipedia. https://en.wikipedia.org/wiki/Near_field_communication

Chapter 2
Ki—The Cordless Kitchen

The Wireless Power Consortium [1] is an open-membership cooperation of 580 companies aiming to create global specifications for wireless power technology. One of the applications of this technology is the cordless kitchen. In the cordless kitchen system, the appliances are powered by inductive power transfer in which a permanently mounted power source containing a coil draws power from the mains and transfers it via electromagnetic induction [2] to a secondary coil placed in the appliance. The power is then converted within the appliance back into electrical power and/or heat for cooking as required.

2.1 Components of Cordless Kitchen

The cordless kitchen mainly consists of three components: power transmitter (or magnetic power source), cordless appliance, and an NFC communication interface, as depicted in Fig. 1.1. Each of these components is described in detail below.

2.1.1 Power Transmitter

A cordless kitchen system consists of one or more inductive or magnetic power sources (MPS) that are integrated into the kitchen countertops or dining tables. The appliances can be powered simply by placing them on top of the MPS. The MPS consists of a coil that draws power from the mains and transfers it using electromagnetic induction to the secondary coil placed in the appliance. This is then converted to electrical power and/or heat within the appliance.

The block diagram of an MPS is shown in Fig. 2.1. It consists of a power transmitter module (refer to Fig. 2.2) which consists of a coil assembly, a communication unit,

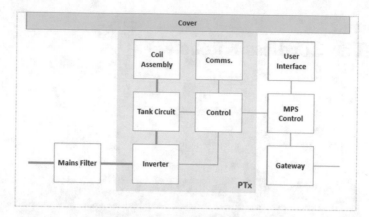

Fig. 2.1 Block diagram of the Magnetic Power Source

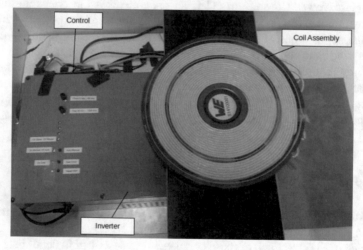

Fig. 2.2 Power Transmitter

a controller, an inverter and a tank circuit. The coil assembly hosts the primary coil for inductive power transfer. The communication unit supports an NFC-based communication using a dedicated antenna that is concentric with the primary coil in order to achieve spatially localized communications with the cordless appliance. Figure 2.3 shows the NFC antenna and Fig. 1.3 shows the antenna assembly in the PTx. The controller is responsible for executing dialogs with the power receiver (PRx), adjusting the operating point of the PTx in response to the requests from the PRx, safeguarding the operating limits of the PTx and handling the Foreign Object Detection (FOD) functionality.

The PTx creates a time-varying magnetic field by driving an alternating current through the primary coil. This magnetic field is captured by the PRx using the secondary coil that is placed in the cordless appliance. If the primary and secondary coils are aligned well after placing the PRx on top of the PTx, they form a loosely coupled transformer. The magnetic coupling is maximum when the two coils are of the same size, are properly aligned and close together (distance and angle between them). The power interface between the PTx and receiver is shown in Fig. 2.4. The cordless kitchen specification allows a positioning tolerance of ± 5 cm, ensuring a power efficiency of at least 90% compared to the equivalent wired appliances.

The PTx implementations typically apply the pulsating rectified mains power to a full- or half-bridge inverter in order to create the driving signal. This results in a driving signal having an envelop as shown in Fig. 2.5. In Fig. 2.5, u_{op} stands for operating voltage and f_{op} for operating frequency.

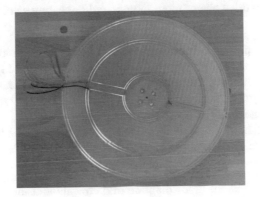

Fig. 2.3 NFC Antenna

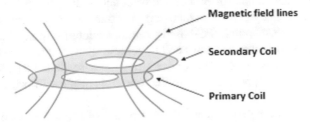

Fig. 2.4 Power interface

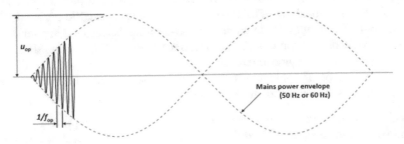

Fig. 2.5 Driving Signal Envelop

Additionally, the MPS also consists of a user interface, a gateway, a mains filter, an MPS controller and a cover. The user interface includes an interface to turn on/off the PTx, display status information, etc. It could also function as a remote user interface to operate the cordless appliance that is being served. The gateway provides Internet connectivity to the MPS and to the cordless appliance (as a proxy) being served.

2.1.2   Cordless Appliance

The cordless kitchen appliances are distinguished into two types. The first type is the motor-driven appliances which convert the magnetic power from the MPS to electrical power to drive one or more motors and/or heating elements. Examples of these are

- Motor-driven appliances: These include appliances with a motor, for example, juicers, mixers and food processors. They consume less than 1 kW of power.
- Appliances with heating: These appliances are designed for heating water or food such as kettles, hot pots, rice cookers and toasters. They may require a power of 2 kW or more due to the presence of heating elements.
- Appliances with heating and motors: These appliances include bread makers, soy milk makers, etc. that combine heating with a mechanical movement for food preparation. They may require more than 1 kW of power.

Figure 2.6 represents the block diagram of a typical motor-driven cordless appliance. These appliances host a power receiver (PRx) that contains a coil assembly, a communication unit, a controller, a tank circuit and a bias supply. The coil assembly hosts the secondary coil for inductive power transfer, as shown in Fig. 1.4. The communication unit supports NFC-based communications using a dedicated antenna that is concentric with the secondary coil to achieve spatially localized communications. The controller is responsible for executing dialogs with the PTx to ensure that the desired amount of power goes into the load. It is also used for driving the user interface.

In addition to these components, the motor-driven appliances also have a user interface, a power converter, a load and a base. The user interface is dependent on the type of the cordless appliance. It typically includes the turn on/off and the PTx wake-up functionalities. The wake-up functionality is used to revive the PTx from a low-power standby mode without re-positioning the cordless appliance. The appliance could offload some of its user interface functionalities to the MPS, and it could also have a remote user interface.

The second type of appliance is the induction heating-based appliance which directly converts the magnetic power from the MPS to heat. For example:

- Cooking appliances: This is a new category of cookware comprising pans, pots and frying pans that would traditionally be used on a stove but that can now be used as a cordless appliance. On these appliances, controls are on the cookware rather than on the stove.

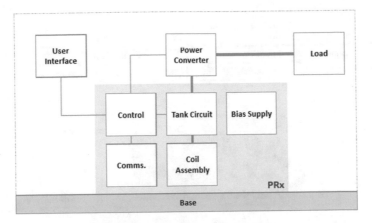

Fig. 2.6 Block diagram of Motor-Driven Cordless Appliances

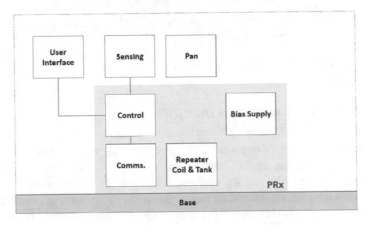

Fig. 2.7 Block diagram of Induction heating-based Cordless Appliances

Figure 2.7 represents the block diagram of a typical induction heating-based cordless appliance. The PRx contained in these appliances has a repeater coil and a tank circuit, a communications unit, a controller and a bias supply. Additionally, these appliances also have a sensing unit, a user interface, a pan and a base. The sensing unit involves one or more sensors monitoring the temperature of the pan or the food contained therein.

Other types of appliances include cooling applications such as wine coolers or portable fans, lighting applications such as moveable luminaries, etc.

Table 2.1 NFC terminology mapping

Cordless Kitchen Specification	ISO/IEC 14443 Standard
Cordless Appliance communication unit	Proximity Card (PICC)
Magnetic Power Source communication unit	Proximity Coupling Device (PCD)

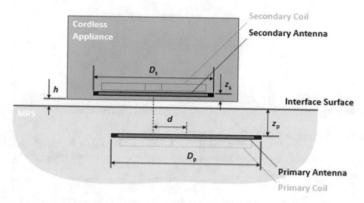

Fig. 2.8 Model of the NFC communication antennas

2.1.3 NFC Communication Interface

The communication interface represents the physical layer for data communications. The cordless kitchen has an NFC channel to establish one-to-one communication between the power source and the cordless appliance. The communication begins as soon as the appliance is placed on top of the MPS. In addition to controlling the amount of power that is transferred, the communication enables smart features such as allowing the MPS to distinguish between cordless appliances and other metallic objects that should not be inadvertently heated.

The communication complies with the ISO/IEC 14443 standard defined for near-field communications. The terminologies used in the cordless kitchen specification are mapped to the ones defined in the ISO/IEC 14443 standard as shown in Table 2.1.

The antennas in the cordless kitchen specification are significantly larger than the ones specified in the ISO/IEC 14443-1 standard. The antennas used in the cordless kitchen are circular in shape and are concentric with the primary and secondary coils used for inductive power transfer (see Figs. 2.3 and 1.4). A model depicting the NFC antennas is shown in Fig. 2.8, and the notations are explained below.

- D_p: Outer diameter of the primary antenna.
- z_p: Distance from the primary antenna to the interface surface.
- D_s: Outer diameter of the secondary antenna.
- z_s: Distance from the secondary antenna to the interface surface.

- d: Alignment tolerance between the axes of the primary and the secondary antennas.
- h: Height of the cordless appliance above the interface surface.

The NFC technology is preferred for the cordless kitchen environment over other wireless technologies like Bluetooth, Zigbee, etc. because of the following advantages of NFC.

- For safety reasons, the cordless kitchen system requires a very short-range detection/communication between the PTx and the appliance. This would ensure that the cordless appliance is detected and the power is transferred only when it is placed on top of the PTx. The most suitable communication technology for such a system would be the NFC, which offers a short-range communication limited to a distance of up to 10 cm, in compliance with the cordless kitchen specification.
- The NFC provides one-to-one communication between devices unlike other short-range wireless protocols. This ensures that a PTx is connected to only one appliance at a time.
- The detection and connection of devices with NFC are almost instantaneous, and the connection happens automatically in a fraction of a second. There is no need to manually set up connections unlike other technologies.
- NFC is capable of supplying power to passive devices alongside the communication, through RF energy harvesting [3]. This would help the appliances to keep some of their functionalities like the User Interface (UI) active even when the PTx is not transferring power.

The cordless kitchen specification defines two profiles of communication: Basic and Extended. The basic profile is based on the NFC Type 2 tag specification and only supports a bit rate of 106 kbps. On the other hand, the extended profile is based on the NFC Type 4 tag specification and supports bit rates of 106, 212, 424 and 848 kbps.

2.2 System Parameters

Figure 2.8 also shows the mechanical model depicting the power transfer in a cordless kitchen. It consists of a magnetic power source hosting the primary coil, cordless appliance hosting the secondary coil and an interface surface between them. The values of the parameters are given in Table 2.2.

The parameter values show that the distance from the secondary coil to the interface surface (z_s) can be greater than the distance of the cordless appliance above the interface surface (h). This indicates that the user can lift the cordless appliance from the interface surface during the power transfer but only up to 1 cm high. Alternatively, the user can insert a magnetically inert protective mat underneath the cordless appliance before starting the power transfer.

Table 2.2   Values of the cordless kitchen parameters

Notation	Value range (mm)
D_p	170–220
z_p	5–30
D_s	80–240 mm (depends on the type of the cordless appliance)
z_s	5–10
d	0–50
h	0–10

The model in Fig. 2.8 and the parameter values also illustrate that the user does not have to precisely align the cordless appliance to the primary coil to operate the power transfer system. The maximum values of alignment tolerance (d) and height (h) have been intertwined for user safety. d is the largest when h is zero, i.e. when the cordless appliance rests on the interface surface, which is typically the case during the normal operation of the power transfer system. d is near zero close to the largest h at which the power transfer system can operate as intended. A value of h greater than zero can occur in practice if a user inserts, for example, a protective mat in between the cordless appliance and the interface surface.

While the model described in this section is based on a power transmitter that uses a single primary coil, practical power transmitter implementations may employ more sophisticated coil designs having mechanical properties that deviate from those used in the system model. One example of such a design is a power transmitter that uses a segmented primary coil. The advantage of this type of coil is that its diameter can be adapted by activating an appropriate number of concentric coil segments. This approach enables the primary coil diameter to be matched to the secondary coil diameter, which results in better efficiency, reduced exposure of the user to the magnetic field and reduced probability that foreign objects prevent the system from operating as intended among other benefits.

Another example of an alternative design is a power transmitter that uses an array of (small) coils to offer the user greater positioning freedom when placing a cordless appliance on the interface surface. When configuring the power transfer, the power transmitter selects the coil or a set of coils that yield the optimum power transfer in terms of efficiency, user exposure to the magnetic field, etc.

2.3   Foreign Object Detection

A potential problem that could occur with wireless power transfer is that power may unintentionally be transferred to foreign metallic objects that are in the vicinity of the power transmitter. For example, if objects like spoons, knives, etc. are placed on top of the power transmitter, the magnetic flux generated by the coil of the transmitter introduces eddy currents in the metallic objects, thus heating them up significantly. This would become a safety hazard. In order to avoid such a scenario, foreign object detection (FOD) mechanism is introduced in the cordless kitchen standard.

An FOD mechanism designed for Ki is described by the inventors in [4]. This mechanism is more accurate and an improved version than the one used in the Qi standard. This is necessary because Ki deals with much larger power and variations in the operating conditions. So the FOD algorithms used in Qi tend to be sub-optimal in some cordless kitchen scenarios, resulting in false or missed detection of foreign objects. The components and the working of FOD in Ki are briefly described below.

For the purpose of FOD, the MPS additionally comprises a driver, a test generator, a foreign object detector and an adapter. The driver generates a drive signal for the transmitter coil, which employs a repeating time frame consisting of a power transfer time interval and a foreign object detection time interval. The test generator generates a test drive signal for the test coil during the foreign object detection time interval. The foreign object detector performs the foreign object detection test based on the measured parameters from the test drive signal. The adapter controls the power transmitter to operate in the foreign object detection initialization mode, prior to entering the power transfer phase. In this mode, the foreign object detector determines the preferred value of the signal parameter for the test drive signal, in response to at least the first message received from the power receiver. This message comprises the following information:

- Expected impact of the power receiver on the reference electromagnetic test signal.
- Constraint for the signal parameter for the test drive signal.
- Difference between the current power receiver operating value and the test reference power receiver operating value.

The preferred value is then set during the foreign object detection time interval. The adapter prevents the power transmitter from entering the power transfer phase if the preferred value does not meet the required criteria.

Similar to the MPS, the power receiver additionally comprises a foreign object detection controller, a message transmitter and a power receiver controller. The foreign object detection controller reduces the load of the power receiver during the foreign object detection time interval. The message transmitter transmits the first message to the power transmitter. The power receiver controller controls the power receiver to operate in the foreign object detection initialization mode, in which the power receiver transmits at least one message to the power transmitter to bias the test drive signal toward causing a reference condition at the power receiver.

2.4 Phases of Operation

The operation of the cordless kitchen system is categorized into different phases. These phases are depicted in Fig. 2.9 and explained in detail below.

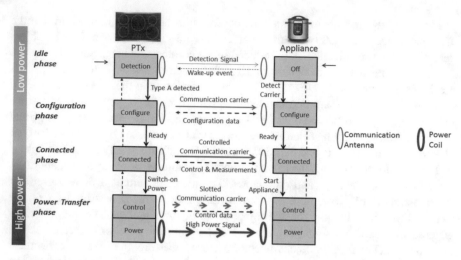

Fig. 2.9 Phases of operation

2.4.1   Idle Phase

In the idle phase, the NFC reader in the PTx is typically polling for the presence of a device in its vicinity, or is in the standby mode waiting for a wake-up event while the appliance is inactive or powered off. The wake-up event is triggered when there is an impedance change at the communication antenna of the PTx or when an event occurs at the PTx's UI or when an Internet event occurs from devices like mobile phones. The PTx consumes very low power in the idle phase. Upon receiving a wake-up event, the PTx detects the RFID device type. Once an NFC Type A device is detected and no other devices are present, the PTx and the appliance enter the configuration phase.

2.4.2   Configuration Phase

In the configuration phase, the PTx collects the configuration data from the appliance, and checks if it is a compliant kitchen device. Two types of configurations take place in this phase. The first one is the NFC configuration, where the device initialization and anti-collision processes are carried out for NFC Type A devices. Bit rate negotiation occurs at this stage. The second type is the static configuration, where the PTx retrieves application-specific data from the appliance. This data includes application identifier, specification version, power class, power rating, etc. Once the configuration is done, the PTx and the appliance enter the connected phase.

2.4.3   Connected Phase

The connected phase mainly concentrates on preparing the system for operation under user control and in the power transfer phase. While the PTx is in the configuration phase, it has full control of the communication and sequence of interactions with the appliance. However, in the connected phase, the PTx delegates this control to the appliance and also enables the appliance to interact with the user.

The PTx and the appliance communicate by exchanging messages carrying information related to power control/negotiation, measurements, state transition requests and auxiliary data. There is no specific order in which these messages are exchanged. The auxiliary data messages are mainly used to connect the appliance to the Internet using the NFC communication channel. This is when the appliance does not have its own Wi-Fi module and relies on PTx for Internet connectivity. These data messages are communicated over the auxiliary channel which is a logical channel over the standard NFC channel. By exchanging these messages, the appliances can have Internet connectivity which can be used for receiving cooking recipes, communicating with a remote user interface, downloading software updates, web browsing, IoT applications, etc.

2.4.4   Power Transfer Phase

In the power transfer phase, the PTx drives current through its power coil to generate a magnetic field from which the appliance retrieves power for its load. The power transfer and the communication in this phase happen in a time-multiplexed fashion as explained earlier. The communication between the PTx and the appliance takes place at the zero crossings of the power signal for a duration of 1.5 ms. There is no specific sequence for the communication, however, the appliance should frequently send heartbeat messages to the PTx to inform about its status. Similarly, the PTx should respond to the control messages of the appliance on time. When the PTx does not hear from the appliance within a predefined interval of time, it cuts off the power and goes to the connected phase. Similar to the connected phase, the PTx and the appliance exchange messages related to power control/negotiation, state transition requests, measurements and auxiliary data.

2.5   NFC Protocol Extensions for the Cordless Kitchen

The idle and configuration phases use the standard NFC protocol, i.e. the standard READ and WRITE commands of the basic profile, and the standard READ BINARY and UPDATE BINARY commands of the extended profile. However, these standard commands cannot be used for the connected and power transfer phases. These phases need dedicated NFC application states and commands to handle the time multiplexing of the NFC communication signals and the power signal.

2.5.1 Dedicated Application States

The NFC communication in the power transfer phase takes place in 1.5 ms time slots that occur periodically (once every 10 ms). During the power transfer, the NFC tag in the appliance loses the detection of the communication carrier. The standard NFC requires the NFC tag to restart from the idle phase upon re-detection of the communication carrier. This would require re-running the NFC configuration which consumes too much time for the given 1.5 ms time slot, so there will not be enough time to exchange other messages. This problem is solved by defining two dedicated application states—Alive and Suspend. When the appliance loses the communication carrier, it goes to the suspend state, and upon re-detection, it does not reconfigure NFC but goes to the alive state where it continues its operation from where it had left off.

2.5.2 Dedicated NFC Commands

New NFC commands are defined to reduce the communication overhead and meet the 1.5 ms time-slot requirement. The READ2 and WRITE2 commands are introduced for the basic profile which is used to read and write data from and to the NFC buffer. Similarly, the READ4 and WRITE4 commands are added to the extended profile. The formats of these commands are explained in detail below (Figs. 2.10 and 2.11).

- READ2:
  Figures 2.10 and 2.11 show the standard READ and READ2 commands, respectively. The standard READ command can carry 16 bytes of data. In the time-multiplexed mode, sending a large amount of data becomes a major constraint.

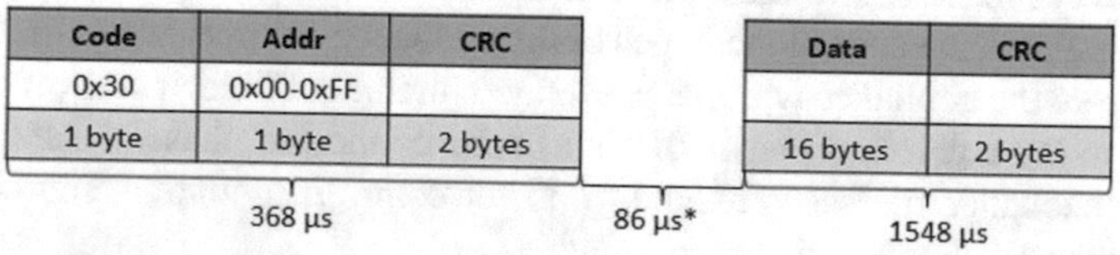

Code	Addr	CRC
0x30	0x00-0xFF	
1 byte	1 byte	2 bytes

368 µs

86 µs*

Data	CRC
16 bytes	2 bytes

1548 µs

Fig. 2.10 READ command in standard NFC

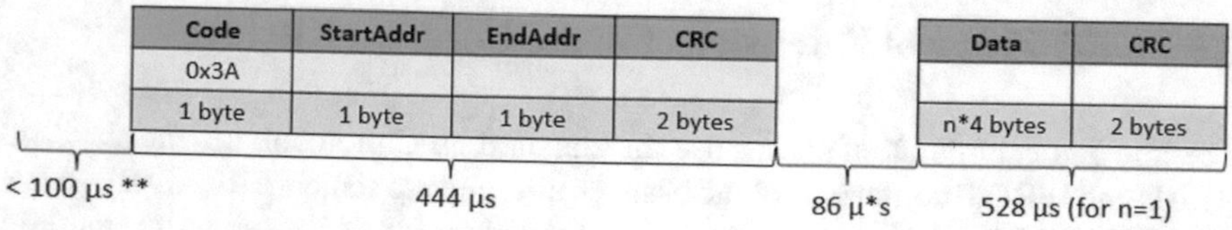

Code	StartAddr	EndAddr	CRC
0x3A			
1 byte	1 byte	1 byte	2 bytes

< 100 µs **

444 µs

86 µ*s

Data	CRC
n*4 bytes	2 bytes

528 µs (for n=1)

Fig. 2.11 READ2 command in dedicated NFC

Numbers of bytes	Over all time
4	100 μs + 444 μs + 86 μs + 520 μs = **1150 μs**
8	100 μs + 444 μs + 86 μs + 860 μs = **1490 μs**
12	100 μs + 444 μs + 86 μs + 1199 μs = **1829 μs**
16	100 μs + 444 μs + 86 μs + 1539 μs = **2169 μs**

Fig. 2.12 Number of bytes that can be sent using READ2 command

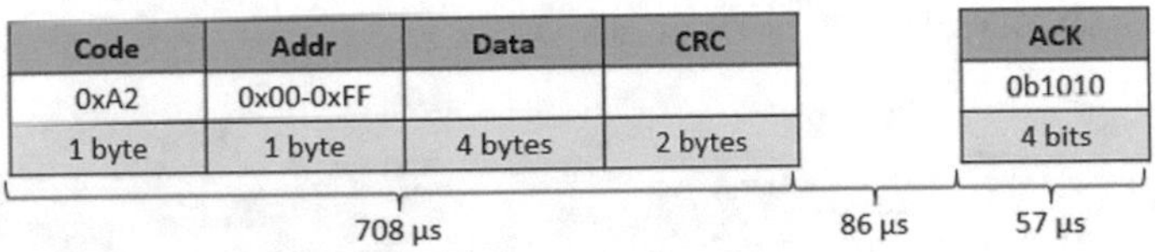

Code	Addr	Data	CRC		ACK
0xA2	0x00-0xFF				0b1010
1 byte	1 byte	4 bytes	2 bytes		4 bits

708 μs   86 μs   57 μs

Fig. 2.13 WRITE command in standard NFC

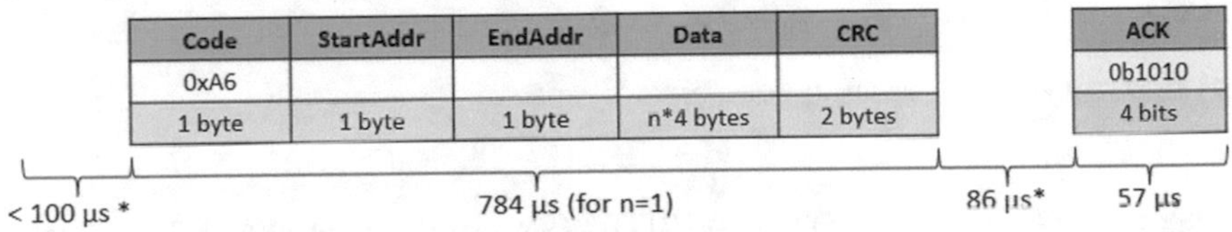

Code	StartAddr	EndAddr	Data	CRC		ACK
0xA6						0b1010
1 byte	1 byte	1 byte	n*4 bytes	2 bytes		4 bits

< 100 μs *   784 μs (for n=1)   86 μs*   57 μs

Fig. 2.14 WRITE2 command in dedicated NFC

Fig. 2.15 Number of bytes that can be sent using WRITE2 command

Numbers of bytes	Over all time
4	100 μs + 784 μs + 86 μs + 57 μs = **1027 μs**
8	100 μs + 1124 μs + 86 μs + 57 μs = **1367 μs**
12	100 μs + 1464 μs + 86 μs + 57 μs = **1707 μs**
16	100 μs + 1804 μs + 86 μs + 57 μs = **2047 μs**

Therefore, the READ2 command is designed to carry data in multiples of 4 bytes. This ensures that the NFC read sequence completes in the given 1.5 ms time slot. The basic NFC protocol uses a bit rate of 106 kbps. The amount of data that can be sent through the NFC channel using the READ2 command within the 1.5 ms time slot is shown in Fig. 2.12. It can be seen that a maximum of 8 bytes (n = 2) of data can fit into a time slot of 1.5 ms.

- WRITE2:
  The standard WRITE command limits its data to 4 bytes only (Fig. 2.13). With the WRITE2 command (Fig. 2.14), more data can be sent in the given 1.5 ms time slot. Similar to the READ2 command, the WRITE2 command carries data in multiples of 4 bytes. From Fig. 2.15, it can be seen that for a bit rate of 106 kbps, 8 bytes (n = 2) of data can be transferred in one time slot of 1.5 ms. The standard READ and WRITE commands are used to read and write the static configuration data in

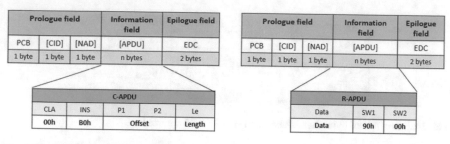

Fig. 2.16 READ_BINARY command in extended NFC

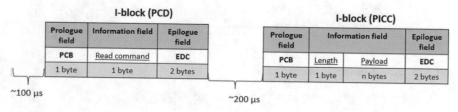

Fig. 2.17 READ4 command in dedicated NFC

the EEPROM. The READ2 and WRITE2 commands are used to read and write data into the dynamic buffer memory of the NFC tag.

- READ4:

The READ_BINARY command of the extended profile is used for reading the static configuration data in the NFC Type 4 tags. Figure 2.16 shows the format of this command. Any amount of data up to a maximum of 255 bytes can be read. However, in the case of time-slotted mode, this command can result in a lot of overhead. Therefore, the new READ4 command is defined with reduced overhead, as shown in Fig. 2.17. This command is used to read the dynamic data from the buffer memory of the Type 4 NFC tag (Fig. 2.18).

Figure 2.18 shows the number of bytes that can be read using the READ4 command for different NFC bit rates. It can be seen that the amount of data that can be transferred increases with the bit rates. Data transfer of up to 104 bytes can be achieved with a bit rate of 848 kbps.

- WRITE4:

Similar to the READ_BINARY command (Fig. 2.19), the UPDATE_BINARY command can carry up to a maximum of 255 bytes of data in its payload. In time-slotted mode, it produces too much overhead because of which the WRITE4 command (Fig. 2.20) is introduced with reduced overhead for an extended profile. The maximum number of bytes that this command can carry in its payload is calculated for different NFC bit rates. The results are shown in Fig. 2.21. With a bit rate of 848 kbps, 103 bytes of data can be transferred across the NFC channel in 1.5 ms.

Bit rate [kbps]	One bit duration [μs]	duration of PCD I-block [μs]	duration of PICC I-block [μs]	FDT [μs]	Start up time [μs]	Over all time [μs]	Number of bytes in payload
106	9,44	358,72	783,52	200	100	1442,24	5
212	4,72	179,36	986,48	200	100	1465,84	19
424	2,36	89,68	1109,2	200	100	1498,88	48
848	1,18	44,84	1149,32	200	100	1494,16	104

Fig. 2.18 Number of bytes that can be sent using READ4 command

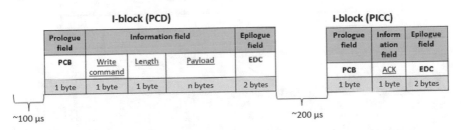

Fig. 2.19 UPDATE_BINARY in extended NFC

Fig. 2.20 WRITE4 command in dedicated NFC

Bit rate [kbps]	One bit duration [μs]	duration of PCD I-block [μs]	duration of PICC I-block [μs]	FDT [μs]	Start up time [μs]	Over all time [μs]	Number of bytes in payload
106	9,44	783,52	358,72	200	100	1442,24	4
212	4,72	986,48	179,36	200	100	1465,84	18
424	2,36	1109,2	89,68	200	100	1498,88	47
848	1,18	1149,32	44,84	200	100	1494,16	103

Fig. 2.21 Number of bytes that can be sent using WRITE4 command

The NFC read and write commands carry messages of the kitchen communication protocol in the connected and power transfer phases. A frame of messages is transferred over one time slot of 1.5 ms. The frame payload contains a header with one or more messages. The message and the frame formats are depicted in Figs. 2.22 and 2.23, respectively. Different types of messages are listed in Fig. 2.24. The messages carry measurements data, operating limits, control data, auxiliary data, etc. and are sent across the NFC channel in connected and power transfer phases.

	b7	b6	b5	b4	b3	b2	b1	b0
Header Byte	Message type (ID)				Message extension (Ext)			
Payload byte	Message payload							
:								
:								
Payload byte								

Fig. 2.22 The kitchen communication protocol: Message format

	b7	b6	b5	b4	b3	b2	b1	b0
B0	RESERVED						FLAGS	
B1	Message Type (ID)				Message extension (Ext)			
:	Message							
:								
:	Message Type (ID)				Message extension (Ext)			
:	Message							
:								
Bn								

Fig. 2.23 The kitchen communication protocol: Frame format

State	Message type	Message extension	Direction	ID	Ext	Meaning
Connected	C-LIMS		Tx→Rx	0x3		Operating limits of transmitter
	C-CTRL		Tx←Rx	0x4		Control during configuration
	C-MEAS	Induced Voltage	Tx←Rx	0x5	0x2	Measurements during configuration
	C-AUXC	Type	Tx←Rx Tx→Rx	0x8		Open/Close auxiliary channel
Power transfer	P-CTRL		Tx←Rx	0x6		Control during power transfer
	P-MEAS	Power Voltage Current Temperature	Tx←Rx	0x7	0x1 0x2 0x3 0x4	Measurements during power transfer
	P-MEAS	Power Current	Tx→Rx	0x7	0x1 0x3	Measurements during power transfer
Connected + Power transfer	X-NULL		Tx←Rx Tx→Rx	0x0		Nothing to send, free response Terminate the data field
	X-RQST		Tx←Rx Tx→Rx	0x1		Request for response, payload = requested message ID
	X-RESP		Tx←Rx Tx→Rx	0x2		Accept/reject of proposal (e.g. operating limits)
	X-AUXD	Type	Tx→Rx Tx←Rx	0x9		Auxiliary data

Fig. 2.24 Message types and extensions in the kitchen communication protocol

References

1. Wikipedia contributors (2021). Wireless Power Consortium. Wikipedia. https://en.wikipedia.org/wiki/Wireless_Power_Consortium
2. Wikipedia contributors (2021). Electromagnetic induction. Wikipedia. https://en.wikipedia.org/wiki/Electromagnetic_induction
3. The Transfer of Power in Near Field Communication (2016). Dummies. https://www.dummies.com/consumer-electronics/transfer-power-near-field-communication/
4. Foreign Object Detection in a Wireless Power Transfer System - Koninklijke Philips N.V. (n.d.) Free Patents Online, https://www.freepatentsonline.com/y2020/0212725.html. Accessed 6 June 2021

Chapter 3
Architectures for Internet Connectivity

In the cordless kitchen, the appliances should be able to connect to the Internet when they are powered, i.e. when they are placed on top of the PTx. It has to be ensured that the appliances maintain the Internet connectivity as long as they are powered on, irrespective of what communication interface is used between the PTx and the appliance. Internet connectivity is not required when they are away from the PTx. One solution to providing connectivity is to install Wi-Fi modules in the appliances. However, there would be drawbacks as mentioned in Chap. 1, which are summarized here:

(a) The Wi-Fi module gets powered only when the appliance is placed on top of the PTx. Powering the appliances with batteries is not desirable as batteries need to be regularly charged and/or replaced.
(b) When the PTx goes into standby, the appliance will be switched off and the Wi-Fi module in the appliance will not be awake to support Internet connectivity. The PTx could supply standby power to the appliance, however, this will not be efficient in terms of power consumption.
(c) The cost of the appliances would increase due to the additional Wi-Fi module and battery.

Hence, having a dedicated Wi-Fi module for every appliance will be unnecessary. To overcome these drawbacks, the Wi-Fi module could be installed in the PTx (or kitchen countertop) instead, and this connection could be shared by all the appliances that use the PTx. To enable this, the existing NFC channel between the appliance and PTx could be used for transmitting the Internet-related information so that the cordless appliances are indirectly connected to the network. Using this solution, the PTx can keep its Wi-Fi module on during standby and wake up the appliance whenever it receives a message for the appliance from a remote user. This would also reduce the cost of the appliance.

Based on this solution, two main architectures can be considered for Internet connectivity: Proxy architecture and Bridge architecture. Both these architectures are to support the TCP/IP protocol. TCP is chosen as the transport layer protocol because

© The Author(s), under exclusive license to Springer Nature Switzerland AG 2021
S. Kashyap et al., *Cook Over IP*,
SpringerBriefs in Applied Sciences and Technology,
https://doi.org/10.1007/978-3-030-85836-0_3

the Internet applications of the cordless kitchen like remote user control, recipe and software uploads require reliable connections. TCP is best suited for such scenarios as it provides a reliable, ordered and error-checked delivery of packets between communicating applications. This chapter explains and evaluates these architectures by discussing their advantages and disadvantages in detail.

3.1   Proxy Architecture

In this architecture, the PTx is installed with a Wi-Fi module or Ethernet connection, so it holds the full TCP/IP stack required for Internet connectivity. The cordless appliance only implements the application layer and sends its application data to the PTx through the NFC channel. The PTx acts like a proxy to the appliance by processing the TCP/IP packets for it, as shown in Fig. 3.1.

In the proxy architecture, the PTx represents the appliance on the Internet. So the TCP session initiation/termination, data packet processing and acknowledgment handling are done by the PTx, as shown in Fig. 3.2. The appliance only sends/receives the application payload. When the PTx receives a TCP/IP packet from the end-user device, it immediately sends out an ACK to the end-user device and then sends the application data to the appliance. It does not wait to check if the application data is delivered correctly to the appliance. Advantages of using this architecture are listed below.

- The implementation of the appliance is simple as it only needs the application layer.
- There is less load on the NFC channel as the appliance sends/receives only the application data. This results in lower system latency.
- Lower cost of the appliance as there is no Wi-Fi module or battery.

However, this architecture has the following disadvantages:

- Reliability is dependent on the PTx implementation because the PTx is responsible for creating and processing the TCP/IP packets of the appliance.

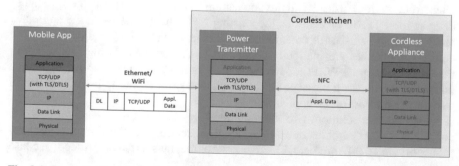

Fig. 3.1 Proxy architecture for Internet connectivity

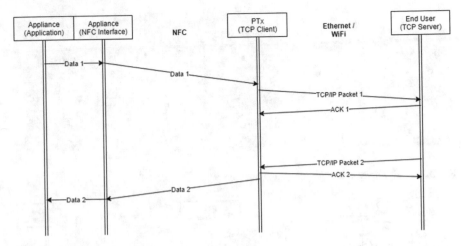

Fig. 3.2 TCP sequence diagram of proxy architecture

- The PTx sends an ACK irrespective of whether the data is delivered to the appliance or not. A special handshake mechanism could be implemented where the PTx waits until the appliance sends an ACK for the data it received. This would increase the latency and also the complexity of implementation.
- The data is not end-to-end protected by the appliance. When the PTx and the appliance are from different manufacturers, the appliance needs to trust the PTx with its application data. There would be possibilities of PTx using the appliance's data for its business purpose without the consent of the user, for example, analyzing user behavior, extracting the appliance's implementation details and sending the packets to a malicious server/user. It is possible to use data encryption techniques to increase the security, however, the PTx would still have the control of processing the TCP/IP packets of the appliance.
- Another disadvantage could be that the PTx manufacturers might not be willing to implement this architecture. There are no advantages for the PTx in this architecture because it only acts like a proxy and has the burden of processing Internet packets for the appliance.
- Appliance is not visible on the network as it does not have its own IP address.

3.2 Bridge Architecture

In this architecture, the PTx contains the Wi-Fi module or Ethernet connection but it acts like a bridge by processing only the data link and the physical layers for the appliance. The higher layers of the TCP/IP stack are implemented by the appliance, as shown in Fig. 3.3. Thus, the appliance does not have to depend on the PTx for TCP/IP packet processing.

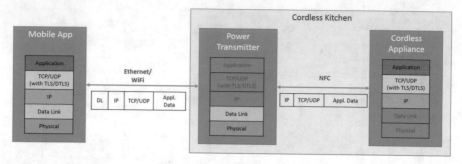

Fig. 3.3 Bridge architecture for Internet connectivity

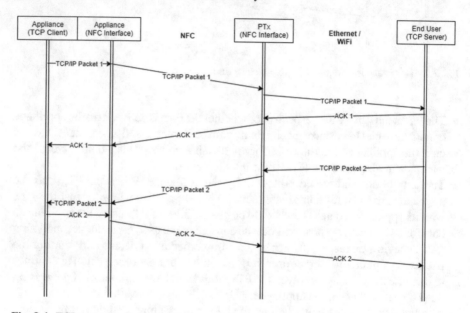

Fig. 3.4 TCP sequence diagram of bridge architecture

Figure 3.4 shows an example TCP sequence diagram using the bridge architecture. In this, the appliance is visible in the network and has a TCP/IP stack of its own. It is completely responsible for TCP session initiation/termination, data packet processing and acknowledgment handling. The PTx merely acts like a bridge by forwarding the appliance's packets to the end-user device. The advantages of using this architecture are as follows:

- The appliance has more control in the process of Internet connectivity. It is only dependent on the PTx for forwarding its TCP/IP packets.
- The data communication can be made more secure by using cryptographic protocols like the Transport Layer Security (TLS) in the appliance stack to ensure data privacy.

- The burden on the PTx is less as it does not have to process the TCP/IP packets for the appliance.
- The appliance will be visible on the network as it will have its own IP address.

  Some of the disadvantages of this architecture are listed below.

- The load on the NFC channel increases due to the overhead introduced by the TCP/IP protocol. This will have a large impact on the latency of the system. As the Internet applications of the cordless kitchen are soft and firm real time, it is very important to have minimal latency and a good response time in the applications. Packet compression techniques could be employed to reduce the latency.
- The implementation of the appliance would be complex due to packet processing and tunneling of the TCP/IP protocol over the NFC channel.

3.3   Comparison of Transmission Latency

The size of the application data in the cordless kitchen depends on the kitchen UI protocol being used. A proprietary protocol called Digital Innovation Communications (DICOMM) Protocol is used for the experiments. The approximate message sizes in the JavaScript Object Notation (JSON)-based variant and the Binary variant of the protocol are shown in Table 3.1.

Figure 3.5 shows the latencies of data exchange using the proxy and bridge architectures at an NFC bit rate of 83.2 kbps in the time-slotted mode. A TCP session exchanging a single data packet is considered for the bridge architecture, and the 6LoWPAN header compression results given in [1] are used. It can be seen that without compression, the latency in the bridge architecture is around 170 ms higher than that of the proxy architecture. This is because of the overhead introduced by the TCP/IP protocol. This overhead remains constant for all data sizes, so it will be less significant at higher sizes. For a data size of 1024 bytes, the latency with the bridge architecture is about 44.6% more than that of the proxy architecture, and with header compression this difference is reduced to about 36.6%. It is possible to have long TCP sessions equal to the duration of the cooking session in order to

Table 3.1 Internet application message sizes using the DICOMM UI protocol

Message type	Message size (Bytes)	
	JSON protocol variant	Binary protocol variant
Switch on/off, Set time/temperature, Keep warm on/off, etc.	30–100	10–35
Status information/Notification	250–300	75–100
Recipe upload	350–1000	125–350

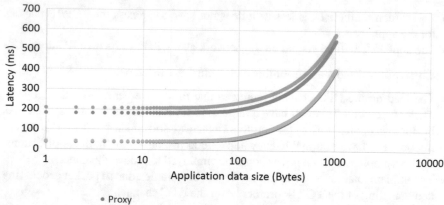

Fig. 3.5 Transmission latency for different application data sizes using proxy and bridge architectures

avoid executing the TCP handshake and termination procedures frequently. In such scenarios the latency obtained with header compression will be very close to that of the proxy architecture as seen in Fig. 3.5.

Reference

1. J. Hui, P. Thubert, Compression Format for IPv6 Datagrams over IEEE 802.15.4-Based Networks. RFC **6282**, 1–24 (2011)

Chapter 4
State of the Art

With the advent of the Internet of Things (IoT) concept, a lot of research is being made to find ways of establishing connectivity between diverse and distinct objects that are based on dissimilar technologies. NFC is a widely used technology which has gained a lot of popularity because of its features like increased security, instant connection procedure, low power, low cost, etc. Over the last few years, there has been some research on providing Internet connectivity to NFC-enabled IoT devices. For the ease of comparison and analysis, the methods of enabling connectivity in the related works are broadly classified into the following types:

1. Tunneling standard TCP/IP protocol over NFC.
2. 6LoWPAN adaptation for TCP/IP protocol over NFC.
3. TCP/IP adaptation mechanisms for high delay networks.

4.1 Tunneling Standard TCP/IP Protocol over NFC

One of the most straightforward ways of providing Internet connectivity to a device through the NFC channel is by tunneling the TCP/IP protocol over the channel. Echevarria et al. [1] introduce a concept called WebTag, which enables direct IP-based access to a sensor tag using NFC technology. It supports secure applications by tunneling the TCP/IP traffic over the NFC carrier. In this method, the sensor tags are equipped with a full TCP/IP suite, web server and an NFC communication channel that connects them to the NFC reader device which behaves like a network gateway. The paper briefly discusses the performance issues caused by bandwidth constraints, increased transmission latency due to differing processing speeds in components, memory limitations, etc. To overcome these challenges, it employs a data fragmentation mechanism, where large packets are fragmented/defragmented at the ends of the NFC channel. It also uses the Van Jacobson packet compression technique [2] to further reduce the latency. Although the paper gives an overview of the effects of tunneling the TCP/IP over NFC, it lacks a detailed analysis of the TCP characteristics that affect the system performance. It uses the uIP stack [3] with standard configuration and does not give enough insight on how the stack can be adapted to the NFC technology to achieve the best results. Furthermore, the paper deals with a scenario where full bandwidth of the NFC channel is available for transmission, unlike the cordless kitchen project where only the slotted bandwidth of NFC is used. So the effects of establishing a TCP/IP connection over a discontinuous channel cannot be realized with the results provided by this paper.

Grunberger et al. [4] propose a concept for a test system that can establish a TCP/IP connection over the NFC channel and tunnel all the TCP/IP data through it. The communication protocol stack used in the system is split into two separate blocks that are connected via Bluetooth. The NFC block hosts the physical (NFCIP-1) and Logical Link Control Protocol (LLCP) layers of the NFC stack, and a PC block is used for the higher TCP/IP layers. The system has the full available bandwidth of the NFC channel, however, as it uses the standard NFC protocol the Protocol Data Unit (PDU) of NFC is limited to 255 bytes. To overcome this limitation, a chaining mechanism is used to transfer large TCP/IP packets, similar to [1]. To reduce the latency, the system ensures that data is transmitted over NFC as soon as the channel is free. It does not employ any compression technique as it mainly focuses on analyzing the performance of the system. The paper provides some test results in terms of TCP retransmission rate and measured data rate using three different configurations of the IP Maximum Transmission Unit (MTU) sizes. However, it does not give a detailed analysis or results taking other aspects of the TCP/IP protocol into account. Although the paper highlights some factors affecting the system latency such as slow data processing in the NFC device and long reaction times due to varying processing speeds in components, it fails to discuss the mitigation techniques or ways of modifying the TCP/IP stack to overcome these challenges. It just gives a preliminary analysis to show the advantages and disadvantages of tunneling TCP/IP over NFC devices.

4.2 6LoWPAN Adaptation for TCP/IP Protocol over NFC

A design space and use-case analysis of transferring IPV6 packets over NFC for resource-constrained IoT devices is provided in [5]. The paper recommends using the 6LoWPAN technology [6] for applications involving constrained node networks such as NFC. 6LoWPAN stands for Internet Protocol (IPv6) over Low-Power Wireless Personal Area Networks (LoWPAN). It is a low-power wireless mesh network which allows devices with limited processing capability to connect directly to the Internet using open standards.

The 6Lo group from the Internet Engineering Task Force (IETF) has been working on standardizing the mechanism for transmitting IPv6 over resource-constrained networks such as NFC [7], Bluetooth Low Energy (BLE) [8] and IEEE 802.15.4 [9]. Youn et al. [7] is an Internet draft from the 6Lo working group that describes a method of remodeling the IPV6 stack to include the data link and the physical layers of the NFC stack by adding 6LowPAN functionalities such as neighbor discovery, address auto-configuration, header compression and fragmentation. In this method, the IP packet is transmitted as the PDU of the NFC LLCP layer.

Park et al. [10] propose a 6LoWPAN adaptation protocol for transmitting IPV6 packets over NFC devices. It involves modifying the standard TCP/IP stack to enable IPV6 communication over the NFC channel using 6LoWPAN techniques described in [7]. The paper provides some simulation results in terms of latency and total packet count. It claims that by using the adaptation layer, the NFC device takes negligible additional time for initialization. It compares the number of IPV6 packet transmissions with and without IP header compression and concludes that the reduction in the number of packet transmissions will be higher for large packet sizes. Other aspects of the performance have not been discussed in detail.

As cordless appliances are resource-constrained IoT devices, using the 6LoWPAN adaptation protocol could be beneficial for the performance. But before directly adopting this technique, it is very important to study the cordless kitchen environment involving the time-slotted NFC channel and analyze how the TCP/IP communication behaves when subjected to such a discontinuous and constrained channel. None of the above works gives insight into this aspect. Furthermore, the works do not provide detailed analysis in terms of the behavior of standard handshake, acknowledgment, retransmission and congestion control mechanisms of TCP over a time-multiplexed NFC channel. They do not study the NFC channel characteristics and quantify the performance based on latency, throughput, retransmissions and bandwidth utilization for different NFC bit rates and Bit Error Rates (BER).

Once an analysis on this level is done, the main bottlenecks of tunneling a heavy-weight protocol like TCP/IP over a constrained channel like time-slotted NFC can be realized. And once these limitations are addressed, techniques like 6LoWPAN and Constrained Application Protocol (CoAP) [11] can be used to further improve the system performance.

4.3   TCP/IP Adaptation Mechanisms for High Delay Networks

There has been some research on the performance of TCP over slow links. Benkö et al. [12] provide a detailed end-to-end TCP performance analysis in GPRS networks in terms of round trip delays, throughput, packet loss ratios, etc. It also quantifies the performance improvements with various TCP parameters like maximum segment size, receiver window size, selective acknowledgments and timestamp usage. Some of the theoretical research aspects in this paper could be mapped to the cordless kitchen system, however, the practical results cannot be used or compared with the cordless kitchen as they deal with completely different wireless technologies.

Katabi et al. [13] propose a new protocol called Explicit Control Protocol (XCP) which gives an improved congestion control mechanism in very high bandwidth-delay product networks. The work mainly concentrates on improving the bandwidth utilization and fairness in bandwidth allocation in addition to reducing the standing queue sizes and packet drops. It describes a congestion feedback mechanism that requires additional fields in the protocol header to carry congestion-related information. As the NFC protocol only supports one-to-one communication, techniques related to fairness control are not applicable to the cordless kitchen system. The congestion feedback mechanism could be used, but the additional overhead in the protocol header could increase the latency on the constrained NFC channel.

Dawkins et al. [14] is an Internet draft from the IETF that gives a generalized overview on the performance implications of slow links on the TCP/IP protocol. It recommends header and payload compression techniques to reduce the latency, TCP buffer auto-tuning to avoid packet drops due to buffer overflow conditions and limited transmit algorithm to trigger fast retransmit and fast recovery in case of packet loss. Leung et al. [15], Klein et al. [16] and Fotiadis et al. [17] provide a couple of techniques to improve the throughput of TCP in wireless networks with high delay variability. They mainly discuss spurious retransmissions that occur in such networks and provide mitigation techniques.

All these works provide some aspects of the TCP/IP behavior in high delay, low bandwidth links but none of them studies the performance of a system containing different types of wireless channels with dissimilar characteristics. There has been no research on slotted TCP mechanisms, where TCP/IP is subjected to a discontinuous communication medium. Some of the techniques proposed in the related work could be used in the cordless kitchen once the TCP protocol is appropriately tuned to the system. This research mainly focuses on adapting the standard TCP/IP protocol stack to the time-multiplexed NFC channel by addressing some major performance bottlenecks. It also provides a detailed analysis of the behavior of important TCP mechanisms in such an environment.

References

1. J.J. Echevarria, J. Ruiz-de-Garibay, J. Legarda, M. Álvarez, A. Ayerbe, J.I. Vazquez, WebTag: web browsing into sensor tags over NFC. Sensors 8675–8690 (2012). https://doi.org/10.3390/s120708675
2. V. Jacobson, Compressing TCP/IP headers for low-speed serial links. RFC **1144**, 1–49 (1990)
3. Wikipedia contributors. UIP (micro IP). Wikipedia (2021). https://en.wikipedia.org/wiki/UIP_(micro_IP)
4. S. Grunberger, J. Langer, Analysis and test results of tunneling IP over NFCIP-1, in *2009 First International Workshop on Near Field Communication* (2009). https://doi.org/10.1109/nfc.2009.21
5. Y. Hong, Y. Choi, M. Shin, J. Youn, Analysis of design space and use case in IPv6 over NFC for resource-constrained IoT devices, in *2015 International Conference on Information and Communication Technology Convergence (ICTC)* (2015). https://doi.org/10.1109/ictc.2015.7354725

6. N. Kushalnagar, G. Montenegro, C.P. Schumacher, IPv6 over low-power wireless personal area networks (6LoWPANs): overview, assumptions, problem statement, and goals. RFC **4919**, 1–12 (2007)
7. J. Youn, Y. Hong, D. Kim, J. Choi, Y. Choi, Transmission of IPv6 packets over near field communication (2000)
8. J. Nieminen, T. Savolainen, M. Isomäki, B. Patil, Z. Shelby, C. Gomez, IPv6 over BLUETOOTH(R) low energy. RFC **7668**, 1–21 (2015)
9. G. Montenegro, N. Kushalnagar, J. Hui, D. Culler, Transmission of IPv6 packets over IEEE 802.15.4 Networks. RFC **4944**, 1–30 (2007)
10. J. Park, S. Lee, S. Bouk, D. Kim, Y. Hong, 6LoWPAN adaptation protocol for IPv6 packet transmission over NFC device, in *Seventh International Conference on Ubiquitous and Future Networks* (2015), pp. 541–543
11. Wikipedia contributors. Constrained application protocol. Wikipedia (2021). https://en.wikipedia.org/wiki/Constrained_Application_Protocol
12. P. Benkö, G. Malicskó, A. Veres, A large-scale, passive analysis of end-to-end TCP performance over GPRS. IEEE INFOCOM 2004 **3**, 1882–1892 (2004)
13. D. Katabi, M. Handley, C. Rohrs, Congestion control for high bandwidth-delay product networks. SIGCOMM '02 (2002)
14. S. Dawkins, G. Montenegro, M. Kojo, V. Magret, End-to-end performance implications of slow links. RFC **3150**, 1–17 (2001)
15. K. Leung, T. Klein, C. Mooney, M. Haner, Methods to improve TCP throughput in wireless networks with high delay variability [3G network example], in *IEEE 60th Vehicular Technology Conference, 2004.* VTC2004-Fall, vol. 4 (2004), pp. 3015–3019
16. T. Klein, K. Leung, R. Parkinson, L.G. Samuel, Avoiding spurious TCP timeouts in wireless networks by delay injection, in *IEEE Global Telecommunications Conference, 2004.* GLOBECOM '04, vol. 5 (2004), pp. 2754–2759
17. G.I. Fotiadis, V. Siris, Improving TCP throughput in 802.11 WLANs with high delay variability, in *2005 2nd International Symposium on Wireless Communication Systems* (2005), pp. 555–559

Chapter 5
Adapting TCP for the Bridge Architecture

Some experiments have been designed to understand what parameters of the TCP/IP protocol affect the performance of the cordless kitchen system. These parameters are recognized and adapted to the system appropriately in order to boost the performance of Internet applications.

5.1 Experimental Setup

The experimental setup consists of three Linux-based systems that behave as the cordless appliance, PTx and the end-user device. The Lightweight IP (LwIP) stack [1] is installed on these, where only the required layers of the stack are utilized, as shown in Fig. 3.3. An Ethernet connection is used between the PTx and the end-user device. An NFC communication channel is set up between the PTx and the cordless appliance. The block diagram of the NFC module used in this experiment is illustrated in Fig. 5.1 and the actual hardware setup is shown in Fig. 5.2. It consists of an NFC Reader/Writer (RW) device, an NFC Card Emulator (CE) device and micro-controllers (MCU) connected to each of them as shown in the figures. The NFC devices have the following characteristics.

- They operate at 13.56 MHz and use the ISO/IEC 14443-4 half-duplex transmission protocol.
- They support bit rates of 212 and 424 kbps.
- They are capable of transferring a chunk of 14 bytes (at 212 kbps) and 30 bytes (at 424 kbps) in one time slot of 1.5 ms. So the bandwidth in the NFC time-slotted mode would be 11.2 kbps (at 212 kbps) and 24 kbps (at 424 kbps).
- They require the data chunk to be available at least 2 ms before the occurrence of a communication time slot.
- The distance of about 3 cm is used between the NFC RW and CE devices (see Fig. 5.3).

© The Author(s), under exclusive license to Springer Nature Switzerland AG 2021
S. Kashyap et al., *Cook Over IP*,
SpringerBriefs in Applied Sciences and Technology,
https://doi.org/10.1007/978-3-030-85836-0_5

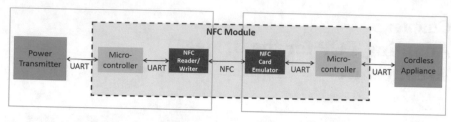

Fig. 5.1 Block diagram of the NFC module

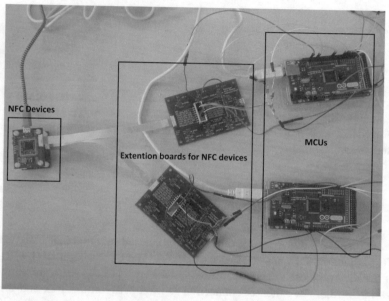

Fig. 5.2 NFC module used for the experiments

- In a kitchen scenario, one may not place the appliance exactly on top of a PTx always. The WPC standard allows a leeway of up to 10 cm and hence requires that no bit errors occur up to this radius from the center of the PTx. Therefore, error correction techniques are not used. The NFC RW device terminates the connection with an NFC CE device when bit errors are detected with the assumption that the appliance is in an unsafe position.

The MCUs used in the module are responsible for the fragmentation of the incoming packets from the TCP/IP stack. The defragmentation is done in the PTx and appliance stacks. The MCUs are also responsible for synchronizing the data transfer with the NFC communication time slots. A serial communication (UART) is used between the MCUs and the NFC devices. According to the cordless kitchen specification, the PTx needs to behave as the NFC RW device and the appliance as the CE device, so the connections are made by interfacing the PTx and the appliance to the appropriate MCUs using UART communication. The MCUs have an incoming packet buffer of

Fig. 5.3 NFC RW and CE modules used in the experiments

Table 5.1 LwIP stack configuration used in the experiments

Configuration type	Value
Protocol Version	IPV4
TCP Maximum Segment Size (MSS)	1024 bytes
Initial Contention Window (CWND) size	4096 bytes
Send buffer size	4096 bytes
Maximum CWND size	8096 bytes
For every ACK received	CWND increases by 1024 bytes
TCP Retransmission Timeout (RTO)	1 s
TCP timer period	500 ms
TCP fast timer period	250 ms

2 kB. The Maximum Transmission Unit (MTU) over the Ethernet channel is 1500 bytes. So a packet buffer size of 2 kB is chosen considering the overheads from the UI protocol and packet processing. They have an interrupt-driven UART reception, and they store and process only one packet at a time.

The proprietary DICOMM UI protocol is used between the end-user device and the appliance. The TCP server and client applications are run on these two devices to exchange data using the UI protocol. The TCP/IP stack configuration used for the experiments is listed in Table 5.1. Table 5.2 summarizes the communication overhead in the cordless kitchen. In the experiments, the cordless appliance is assigned with an IP address of 192.168.1.102, and the end-user device with 192.168.1.202, as shown in Fig. 5.4. Note that for ease of implementation, the PTx is also given the same IP address of 192.168.1.102 as that of the appliance. Wireshark packet analyzer tool is used over the Ethernet link. The packets over the NFC channel cannot be captured by this tool, so logs from the NFC and the TCP/IP stacks are also used for analysis.

Table 5.2 Communication overhead in the cordless kitchen system

Overhead type	Size (Bytes)
IPV4	20
TCP	20
UI protocol	8
Packet handling	8
NFC protocol	4 per time slot
Total	**56 + (4 * No. of time slots per packet)**

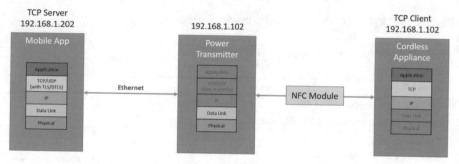

Fig. 5.4 TCP client as the cordless appliance and TCP server as the mobile app

5.2   Challenges in Adapting TCP

TCP is a transport layer protocol which is responsible for ensuring reliable transmission of data across Internet-connected networks. It is called a connection-oriented protocol as it establishes a virtual connection between two hosts using a series of request and reply messages. It divides the messages or files to be transmitted into segments that are encapsulated into the body of the IP packets. Upon reaching the destination, these segments are reassembled to form the complete message or file. TCP defines a parameter known as Maximum Segment Size (MSS) which represents the maximum payload size a TCP segment can hold excluding the TCP header. It is basically the application data size that can be sent in a single TCP/IP packet. TCP executes a three-way handshake sequence for connection establishment between two hosts, as shown in Fig. 5.5. During the handshake, the hosts agree upon the MSS value that will be used during the data transfer. Once the hosts finish exchanging data, the TCP session will be terminated using the connection termination procedure.

While the connection is established and the data transfer is in progress, TCP uses several mechanisms such as congestion control and flow control to provide a reliable connection. The congestion control includes the slow start, congestion avoidance, fast retransmit and fast recovery mechanisms [2]. TCP maintains a retransmission timer to detect and retransmit lost segments. Each time a segment is sent, TCP starts the retransmission timer which begins at a predetermined value called Retransmission

Fig. 5.5 TCP connection establishment, data transfer and connection termination procedures

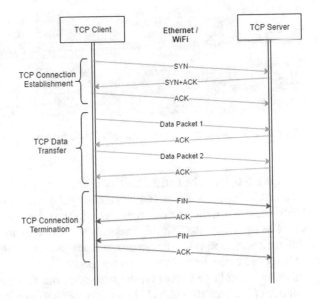

Timeout (RTO) and counts down over time. If this timer expires before an acknowledgment is received for the segment, TCP retransmits the segment assuming that the packet is lost. The RTO value for segments is set dynamically by measuring the Round Trip Time (RTT) of the previous segments. This helps in setting appropriate RTO values by understanding the current delay on the channel.

The flow control determines the rate at which data is transmitted between the sender and receiver in a TCP session. TCP uses a sliding window mechanism for flow control. Due to the limited buffer space, the sender and receiver maintain a congestion window (CWND) and receive window which represent the amount of unacknowledged data that can be in transit at any given time. (Note: Please refer to [2] for a detailed explanation on the working of the TCP/IP protocol).

In this book, the TCP MSS, RTO, RTT and CWND parameters are considered while adapting TCP/IP for the slotted NFC channel as they are the fundamental factors that affect the performance of the cordless kitchen. This chapter mainly concentrates on adapting the TCP RTO and RTT parameters to the given system. The effects of TCP MSS and CWND sizes on performance are discussed in Chap. 7.

To analyze the performance of tunneling TCP/IP over the time-slotted NFC channel, a TCP session is established over NFC at a bit rate of 11.2 kbps and an initial TCP RTO value of 1 s. The PTx and appliance are configured to run the TCP server (192.168.1.202) and client (192.168.1.102) applications, respectively. A payload size equal to TCP MSS of 1024 bytes is exchanged in the session, which generates an NFC payload size of 1080 bytes, including all the overheads mentioned in Table 5.2. The result obtained is depicted in Fig. 5.6. It shows the output from the Wireshark tool taken over the Ethernet link. The packets over the NFC channel are not visible in the capture.

No.	Time	Source	Destination	Protocol	Length	Info
1	0.000000s	MS-NLB-PhysServe_	Broadcast	ARP	42	Who has 192.168.1.202? Tell 192.168.1.102
2	0.000519s	MS-NLB-PhysServe_	MS-NLB-PhysServe_	ARP	60	192.168.1.202 is at 02:12:34:56:78:cd
3	0.000579s	192.168.1.102	192.168.1.202	TCP	58	49153 → 7891 [SYN] Seq=0 Win=8096 Len=0 MSS=1024
4	0.001024s	192.168.1.202	192.168.1.102	TCP	60	7891 → 49153 [SYN, ACK] Seq=0 Ack=1 Win=8096 Len=0 MSS=1024
5	0.139693s	192.168.1.102	192.168.1.202	TCP	54	49153 → 7891 [ACK] Seq=1 Ack=1 Win=8096 Len=0
6	1.860508s	192.168.1.102	192.168.1.202	TCP	1078	49153 → 7891 [PSH, ACK] Seq=1 Ack=1 Win=8096 Len=1024
7	1.861212s	192.168.1.202	192.168.1.102	TCP	1078	7891 → 49153 [PSH, ACK] Seq=1 Ack=1025 Win=8096 Len=1024
8	2.758889s	192.168.1.102	192.168.1.202	TCP	1078	[TCP Retransmission] 7891 → 49153 [PSH, ACK] Seq=1 Ack=1025 Win=8096 Len=1024
9	3.860816s	192.168.1.202	192.168.1.102	TCP	1078	[TCP Spurious Retransmission] 49153 → 7891 [PSH, ACK] Seq=1 Ack=1 Win=8096 Len=1024
10	3.861312s	192.168.1.102	192.168.1.202	TCP	60	[TCP Dup ACK 7#1] 7891 → 49153 [ACK] Seq=1025 Ack=1025 Win=8096 Len=0
11	4.202049s	192.168.1.202	192.168.1.102	TCP	54	49153 → 7891 [FIN, ACK] Seq=1025 Ack=1025 Win=8096 Len=0
12	4.202545s	192.168.1.102	192.168.1.202	TCP	60	7891 → 49153 [ACK] Seq=1025 Ack=1026 Win=8095 Len=0
13	4.202651s	192.168.1.102	192.168.1.202	TCP	60	7891 → 49153 [FIN, ACK] Seq=1025 Ack=1026 Win=8095 Len=0
14	4.412743s	192.168.1.102	192.168.1.202	TCP	54	49153 → 7891 [ACK] Seq=1026 Ack=1026 Win=8095 Len=0

Fig. 5.6 TCP session with a data exchange of 1080 bytes at 11.2 kbps

In Fig. 5.6, it can be noticed that there are some retransmitted and duplicate ACK (Dup ACK) packets in the TCP session (highlighted in black). The Dup ACKs are transmitted when the receiver sees a gap in the sequence number of received packets. The logs from the TCP/IP stacks show that there are two retransmissions from the appliance, and one retransmission from the PTx followed by a Dup ACK sent in response to the retransmission from the appliance. The presence of such retransmissions has a large impact on the latency of the TCP session. This is because the latency of the system is already in the order of seconds due to the constrained bandwidth of the NFC channel, and transmitting these extra packets would increase the latency even further, impacting the end-user experience. It is therefore important to eliminate these retransmissions by identifying the cause of their occurrence. The subsequent sections discuss in detail the classification of these retransmissions and their elimination techniques.

5.2.1 TCP Spurious Retransmissions

The packets 8 and 9 in Fig. 5.6 are spurious retransmissions from the client and server stacks, respectively. Spurious retransmissions occur when the sender thinks that its packet is lost and sends it again, even though the receiver sent an acknowledgment for it. This happens when the sender experiences a timeout before the ACK is received due to the TCP RTO value being very small compared to that of the packet RTT. Figure 5.7 depicts a case where a spurious retransmission problem occurs. Here, the appliance stack does not wait long enough to receive the ACK from the end-user device, which leads to a series of unnecessary transmissions.

To confirm if some or all of these are spurious retransmissions, the experiment is repeated with smaller NFC payload sizes. Table 5.3 gives an overview of the number of retransmissions and Dup ACKs observed for different payload sizes at a bit rate of 11.2 kbps. It can be noticed that as the data size decreases, the number of retransmissions also decreases. If these are spurious retransmissions, this behavior makes sense because smaller data sizes will have smaller RTT. So the chances of the RTO timer of 1 s getting triggered will be less which would result in fewer or no spurious retransmissions. At 11.2 kbps, the RTT of a 500-byte packet is about 1.1 s,

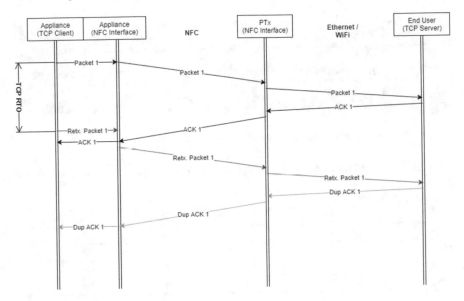

Fig. 5.7 Spurious retransmissions in a TCP session

Table 5.3 Number of retransmissions and Dup ACKs in TCP sessions for varying payload sizes at 11.2 kbps

Payload on NFC (Bytes)	Appliance		PTx	
	Retxs.	Dup ACKs	Retxs.	Dup ACKs
250	1	0	0	0
500	1	0	1	0
1000	2	0	1	1
1080	2	0	1	1

resulting in a total of two retransmissions, and the RTT of a 250-byte packet is about 0.6 s, which results in only a single retransmission.

The experiment is repeated at an NFC bit rate of 24 kbps for a payload size of 1080 bytes. The result is depicted in Fig. 5.8. It can be seen that there is one retransmission from both PTx and appliance, and two Dup ACKs only from the appliance stack. At higher bit rates, the RTT of the packets over NFC will be even less. This would further reduce the number of spurious retransmissions. Table 5.4 summarizes the results for different NFC payload sizes exchanged in the TCP session at 24 kbps. It can be observed that fewer retransmissions are observed compared to that in 11.2 kbps.

These experiments confirm that the TCP RTO value is very small for the given system which makes the stacks timeout sooner than the expected arrival time of the acknowledgment, leading to spurious retransmissions. To overcome this, the TCP packet size could be reduced such that its RTT will be less than the RTO value that is set by default. For this, however, the TCP MSS value will have to be reduced, which

No.	Time	Source	Destination	Protocol	Length	Info
1	0.000000s	MS-NLB-PhysServe…	Broadcast	ARP	42	Who has 192.168.1.202? Tell 192.168.1.102
2	0.000606s	MS-NLB-PhysServe…	MS-NLB-PhysServe…	ARP	60	192.168.1.202 is at 02:12:34:56:78:cd
3	0.000726s	192.168.1.102	192.168.1.202	TCP	58	49153 → 7891 [SYN] Seq=0 Win=8096 Len=0 MSS=1024
4	0.001134s	192.168.1.202	192.168.1.102	TCP	60	7891 → 49153 [SYN, ACK] Seq=0 Ack=1 Win=8096 Len=0 MSS=1024
5	0.110069s	192.168.1.102	192.168.1.202	TCP	54	49153 → 7891 [ACK] Seq=1 Ack=1 Win=8096 Len=0
6	1.227402s	192.168.1.102	192.168.1.202	TCP	1078	49153 → 7891 [PSH, ACK] Seq=1 Ack=1 Win=8096 Len=1024
7	1.228032s	192.168.1.202	192.168.1.102	TCP	1078	7891 → 49153 [PSH, ACK] Seq=1 Ack=1025 Win=8096 Len=1024
8	1.755006s	192.168.1.102	192.168.1.202	TCP	1078	[TCP Retransmission] 7891 → 49153 [PSH, ACK] Seq=1 Ack=1025 Win=8096 Len=1024
9	1.790349s	192.168.1.102	192.168.1.202	TCP	54	49153 → 7891 [FIN, ACK] Seq=1025 Ack=1025 Win=8096 Len=0
10	1.790827s	192.168.1.202	192.168.1.102	TCP	60	7891 → 49153 [ACK] Seq=1025 Ack=1026 Win=8095 Len=0
11	1.790990s	192.168.1.202	192.168.1.102	TCP	60	7891 → 49153 [ACK] Seq=1025 Ack=1026 Win=8095 Len=0
12	2.365291s	192.168.1.102	192.168.1.202	TCP	54	[TCP Dup ACK 9#1] 49153 → 7891 [ACK] Seq=1026 Ack=1025 Win=8096 Len=0
13	2.755409s	192.168.1.202	192.168.1.102	TCP	60	[TCP Spurious Retransmission] 7891 → 49153 [FIN, ACK] Seq=1025 Ack=1026 Win=8095 Len=0
14	2.837292s	192.168.1.102	192.168.1.202	TCP	54	49153 → 7891 [ACK] Seq=1026 Ack=1026 Win=8095 Len=0

Fig. 5.8 TCP session with a data exchange of 1080 bytes at 24 kbps

Table 5.4 Number of retransmissions and Dup ACKs in TCP sessions for varying payload sizes at 24 kbps

Payload on NFC (Bytes)	Appliance		PTx	
	Retxs.	DUP ACKs	Retxs.	DUP ACKs
250	1	0	0	0
500	1	0	0	0
1000	1	0	0	0
1080	1	2	1	0

would lower the goodput of the system. Therefore, it makes more sense to adjust the RTO value appropriately to suit the system.

The TCP/IP stack updates the RTO for its packets dynamically by constantly measuring the RTT of its data packets. The authors of [3, 4] propose methods of avoiding spurious retransmissions in wireless networks that have high delay variability by injecting delays into the network. These delays increase the RTT of the packets and hence the calculated TCP RTO values. Leung et al. [5] present another technique to increase the TCP RTO value by increasing the mean deviation of the measured packet RTT. Although these solutions promise to reduce spurious retransmissions, they will not be very useful in the cordless kitchen system because TCP timeout occurs for the first data packet of the TCP session, for which the RTT measurement has not been made yet. The stack therefore ends up using the initial TCP RTO that is set at compile time, for this packet. Moreover, the TCP sessions in this system can be short, so there will not be enough time to adapt to the dynamically calculated RTO values. Furthermore, this system uses low bandwidth and large delay channel; it is therefore necessary to remove the retransmissions as much as possible, right from the beginning of the TCP session, to ensure a good end-user experience.

To avoid spurious retransmissions, the initial TCP RTO needs to be greater than the RTT of the maximum packet size traveling through the NFC channel. This value gets automatically updated after TCP starts making RTT measurements. It is recommended to set the RTO slightly higher than the RTT of the data packet. This guarantees that there are no spurious retransmissions and also ensures quick retransmission in case of packet loss. Figure 5.9 shows a scenario where the appliance stack waits sufficiently to receive an ACK for the transmitted data packet. It can be seen

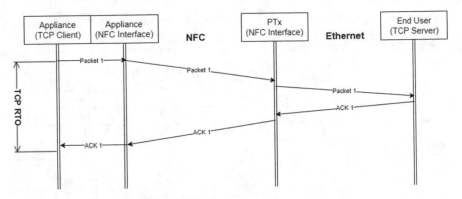

Fig. 5.9 Spurious retransmissions solved by increasing the initial RTO value

No.	Time	Source	Destination	Protocol	Length	Info
1	0.000000s	MS-NLB-PhysServe…	Broadcast	ARP	42	Who has 192.168.1.202? Tell 192.168.1.102
2	0.000513s	MS-NLB-PhysServe…	MS-NLB-PhysServe…	ARP	60	192.168.1.202 is at 02:12:34:56:78:cd
3	0.000630s	192.168.1.102	192.168.1.202	TCP	58	49153 → 7891 [SYN] Seq=0 Win=8096 Len=0 MSS=1024
4	0.001072s	192.168.1.202	192.168.1.102	TCP	60	7891 → 49153 [SYN, ACK] Seq=0 Ack=1 Win=8096 Len=0 MSS=1024
5	0.139762s	192.168.1.102	192.168.1.202	TCP	54	49153 → 7891 [ACK] Seq=1 Ack=1 Win=8096 Len=0
6	5.872002s	192.168.1.102	192.168.1.202	TCP	1078	49153 → 7891 [PSH, ACK] Seq=1 Ack=1 Win=8096 Len=1024
7	5.872726s	192.168.1.202	192.168.1.102	TCP	1078	7891 → 49153 [PSH, ACK] Seq=1 Ack=1025 Win=8096 Len=1024
8	7.126788s	192.168.1.102	192.168.1.202	TCP	54	49153 → 7891 [FIN, ACK] Seq=1025 Ack=1025 Win=8096 Len=0
9	7.127341s	192.168.1.202	192.168.1.102	TCP	60	7891 → 49153 [ACK] Seq=1025 Ack=1026 Win=8095 Len=0
10	7.127436s	192.168.1.202	192.168.1.102	TCP	60	7891 → 49153 [FIN, ACK] Seq=1025 Ack=1026 Win=8095 Len=0
11	7.337531s	192.168.1.102	192.168.1.202	TCP	54	49153 → 7891 [ACK] Seq=1026 Ack=1026 Win=8095 Len=0

Fig. 5.10 TCP session with a data exchange of 1080 bytes at 11.2 kbps with initial RTO of 5 s

that eliminating the retransmissions and Dup ACKs saves a lot of time by reducing the number of packets transmitted over the constrained NFC channel, which in turn improves the overall responsiveness of the system.

In the presence of so many retransmissions, it is difficult to estimate and generalize the exact RTT of the packets. Therefore, initially a high RTO value greater than the total TCP session duration is chosen so that all the spurious retransmissions are eliminated, which would make the analysis of the packet RTT easier. At an NFC bit rate of 11.2 kbps, the average TCP session duration with 1080 bytes of data exchange is about 4.56 s, so an initial RTO of 5 s is chosen for both the client and server stacks to make sure that TCP does not timeout before the first data acknowledgment is received.

Figure 5.10 shows the result after updating the initial TCP RTO to 5 s. No spurious retransmissions are observed in the TCP session, and both server and client stacks wait sufficiently to receive an acknowledgment. However, there is one retransmission at the appliance as indicated by the stack logs. It can be seen that the time difference between packets 5 and 6 is about 5 s, which is equal to the RTO set. This implies that packet 6 is a retransmitted packet. The fact that this was not removed by setting a high RTO suggests that this is not a spurious retransmission. Further analysis on this will be discussed in Sect. 5.2.2.

Similar results are obtained at other data sizes (refer Table 5.5), where one retransmission exists from the appliance stack. The experiment is repeated at 24 kbps by

Table 5.5 Number of retransmissions and Dup ACKs in TCP sessions for varying payload sizes at 11.2 kbps with an initial TCP RTO of 5 s

Payload on NFC (Bytes)	Appliance		PTx	
	Retxs.	DUP ACKs	Retxs.	DUP ACKs
250	1	0	0	0
500	1	0	0	0
1000	1	0	0	0
1080	1	0	0	0

setting an initial RTO of 3 s, as the average TCP session duration is around 2.45 s. It is again observed that although all spurious retransmissions are removed, one retransmission from the appliance stack still exists.

It is very important to set an optimum TCP RTO value for every packet to avoid spurious retransmissions. For this, it is necessary to understand the channel TCP is dealing with. Using RTT of the previous packets cannot be the only factor that should be considered for estimating the RTO for the data packets. The estimation needs to be done by analyzing the following parameters as well.

- NFC bit rate being used;
- Speed/bandwidth of the channel between PTx and the end-user device;
- Total packet size, as the RTT depends on the size of the packet.

Note: For the initial TCP RTO, the RTT of the maximum possible packet size that can be transferred over the NFC channel should be used.

Generalizing the RTT estimation procedure would be more precise when the TCP session is free from all kinds of retransmissions and Dup ACKs. Therefore, it is necessary to first eliminate the remaining retransmissions before proceeding to a generalized approach for setting an optimum TCP RTO, which is explained in Sect. 5.3.2.1.

5.2.2 Packet Drops Due to Small Inter-Packet Delay

In Fig. 5.10, although the spurious retransmissions and duplicate ACKs are removed, the total time of the TCP connection has considerably increased to about 7.4 s compared to the one with an initial RTO of 1 s, which was 4.56 s on an average. This sudden increase takes place between packets 5 and 6 (highlighted in Fig. 5.10). The time difference of about 5 s between these packets, which is equal to the initial RTO set, suggests that packet 6 is a retransmitted packet from the appliance.

The result of the experiment at a bit rate of 24 kbps with an exchange of 1080 bytes of NFC payload and an initial RTO of 3 s is represented in Fig. 5.11. Again,

No.	Time	Source	Destination	Protocol	Length Info
1	0.000000s	MS-NLB-PhysServe…	Broadcast	ARP	42 Who has 192.168.1.202? Tell 192.168.1.102
2	0.000491s	MS-NLB-PhysServe…	MS-NLB-PhysServe…	ARP	60 192.168.1.202 is at 02:12:34:56:78:cd
3	0.000565s	192.168.1.102	192.168.1.202	TCP	58 49153 → 7891 [SYN] Seq=0 Win=8096 Len=0 MSS=1024
4	0.001010s	192.168.1.202	192.168.1.102	TCP	60 7891 → 49153 [SYN, ACK] Seq=0 Ack=1 Win=8096 Len=0 MSS=1024
5	0.078918s	192.168.1.102	192.168.1.202	TCP	54 49153 → 7891 [ACK] Seq=1 Ack=1 Win=8096 Len=0
6	3.226806s	192.168.1.102	192.168.1.202	TCP	1078 49153 → 7891 [PSH, ACK] Seq=1 Ack=1 Win=8096 Len=1024
7	3.227540s	192.168.1.202	192.168.1.102	TCP	1078 7891 → 49153 [PSH, ACK] Seq=1 Ack=1025 Win=8096 Len=1024
8	3.789701s	192.168.1.102	192.168.1.202	TCP	54 49153 → 7891 [FIN, ACK] Seq=1025 Ack=1025 Win=8096 Len=0
9	3.790039s	192.168.1.202	192.168.1.102	TCP	60 7891 → 49153 [ACK] Seq=1025 Ack=1026 Win=8095 Len=0
10	3.790048s	192.168.1.202	192.168.1.102	TCP	60 7891 → 49153 [FIN, ACK] Seq=1025 Ack=1026 Win=8095 Len=0
11	3.913538s	192.168.1.102	192.168.1.202	TCP	54 49153 → 7891 [ACK] Seq=1026 Ack=1026 Win=8095 Len=0

Fig. 5.11 TCP session with a data exchange of 1080 bytes at 24 kbps with initial RTO of 3 s

No.	Time	Source	Destination	Protocol	Length Info
1	0.000000s	MS-NLB-PhysServe…	Broadcast	ARP	42 Who has 192.168.1.202? Tell 192.168.1.102
2	0.000491s	MS-NLB-PhysServe…	MS-NLB-PhysServe…	ARP	60 192.168.1.202 is at 02:12:34:56:78:cd
3	0.000606s	192.168.1.102	192.168.1.202	TCP	58 49153 → 7891 [SYN] Seq=0 Win=8096 Len=0 MSS=1024
4	0.001128s	192.168.1.202	192.168.1.102	TCP	60 7891 → 49153 [SYN, ACK] Seq=0 Ack=1 Win=8096 Len=0 MSS=1024
5	0.001253s	192.168.1.102	192.168.1.202	TCP	54 49153 → 7891 [ACK] Seq=1 Ack=1 Win=8096 Len=0
6	0.001338s	192.168.1.102	192.168.1.202	TCP	1078 49153 → 7891 [PSH, ACK] Seq=1 Ack=1 Win=8096 Len=1024
7	0.002062s	192.168.1.202	192.168.1.102	TCP	1078 7891 → 49153 [PSH, ACK] Seq=1 Ack=1025 Win=8096 Len=1024
8	0.002302s	192.168.1.102	192.168.1.202	TCP	54 49153 → 7891 [FIN, ACK] Seq=1025 Ack=1025 Win=8096 Len=0
9	0.002657s	192.168.1.202	192.168.1.102	TCP	60 7891 → 49153 [ACK] Seq=1025 Ack=1026 Win=8095 Len=0
10	0.002678s	192.168.1.202	192.168.1.102	TCP	60 7891 → 49153 [FIN, ACK] Seq=1025 Ack=1026 Win=8095 Len=0
11	0.002801s	192.168.1.102	192.168.1.202	TCP	54 49153 → 7891 [ACK] Seq=1026 Ack=1026 Win=8095 Len=0

Fig. 5.12 TCP session with a data exchange of 1080 bytes without the NFC channel

the time difference of about 3 s between packets 5 and 6, equal to the initial RTO, suggests that packet 6 has been retransmitted by the appliance stack, just like the previous case. The NFC interface and appliance stack logs reveal that the first data packet (packet 6) which was sent right after packet 5 was dropped at the interface by the NFC module. This is the reason that it cannot be seen on the Wireshark capture.

To understand why the packet was dropped, it is important to study the time delay between TCP/IP packets exchanged between two devices in normal situations, i.e. without the NFC channel. This would give an idea of what the ideal delay between packets 5 and 6 should have been. Figure 5.12 shows the packet capture taken between two devices connected via Ethernet. The TCP client is at 192.168.1.102 and the TCP server is at 192.168.1.202.

The average delay between packets 5 and 6, representing the ACK of the TCP handshake and the first data packet, respectively, is around 50.6 µs. This means that when the NFC channel is being used, the TCP stack generates packet 6 50.6 µs after packet 5 and pushes it to the NFC channel. This inter-packet delay between consecutive packets would be too small for the NFC channel as it is half-duplex and can only transmit packets one at a time. Moreover, the NFC module used in this setup can store and process only a single packet at a time. It discards all the packets that it receives while it is transmitting. In this case, packet 5 takes around 69.04 ms to travel through the NFC channel at 11.2 kbps and around 39.76 ms at 24 kbps. So when the appliance stack sends packet 6 only 50.6 µs after sending packet 5, the NFC module discards it as it will be busy transmitting packet 5. Figure 5.13 summarizes this problem. It shows how a small inter-packet delay between two consecutive packets causes packet loss, which impacts the overall latency of the TCP session.

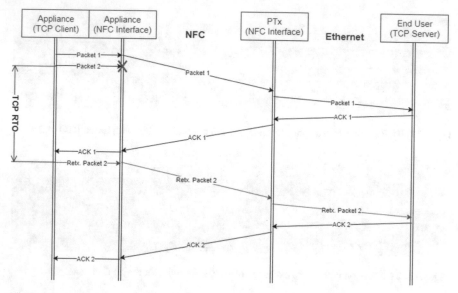

Fig. 5.13 Packet drop at the NFC interface due to small inter-packet delay

5.3   Addressing the Challenges

Both the outlined problems are due to the low data rate, time-slotted NFC channel. The packet drops at the NFC interface causes retransmissions, and spurious retransmissions can cause further packet drops at the NFC interfaces. We break this tie by first solving the packet drops issue, and then address the spurious retransmissions problem.

5.3.1   Avoiding Packet Drops Due to Small Inter-Packet Delay

To avoid packet drops at the NFC interface caused due to small inter-packet delay between consecutive packets, there must be a way for the stack to sense the NFC channel before sending packets to it. An NFC channel sensing mechanism is implemented where the NFC channel notifies the stack when it is busy or free. The stack keeps track of this and sends the packets only when the channel is free. By implementing this mechanism on both the ends of the NFC channel, i.e. in the NFC-appliance and the NFC-PTx interfaces, it can be ensured that packet drops are not caused due to sending the packets too soon into the channel. This way the processing speed of the TCP/IP stack can be brought down to match the speed of the NFC channel so that they function in sync. Figure 5.14 shows how the packet drop problem is solved by implementing the NFC channel sensing mechanism.

Figures 5.15 and 5.16 show the result after implementing the mechanism at 11.2 kbps and 24 kbps, respectively. The TCP sessions are free from retransmis-

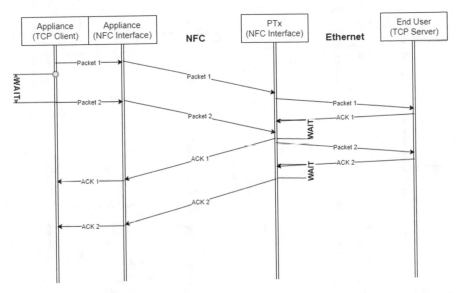

Fig. 5.14 NFC channel sensing mechanism

No.	Time	Source	Destination	Protocol	Length	Info
1	0.000000s	MS_NLB-PhysServe…	Broadcast	ARP	42	Who has 192.168.1.202? Tell 192.168.1.102
2	0.000403s	MS-NLB-PhysServe…	MS-NLB-PhysServe…	ARP	60	192.168.1.202 is at 02:12:34:56:78:cd
3	0.000449s	192.168.1.102	192.168.1.202	TCP	58	49153 → 7891 [SYN] Seq=0 Win=8096 Len=0 MSS=1024
4	0.000803s	192.168.1.202	192.168.1.102	TCP	60	7891 → 49153 [SYN, ACK] Seq=0 Ack=1 Win=8096 Len=0 MSS=1024
5	0.139703s	192.168.1.102	192.168.1.202	TCP	54	49153 → 7891 [ACK] Seq=1 Ack=1 Win=8096 Len=0
6	1.325267s	192.168.1.102	192.168.1.202	TCP	1078	49153 → 7891 [PSH, ACK] Seq=1 Ack=1 Win=8096 Len=1024
7	1.325804s	192.168.1.202	192.168.1.102	TCP	1078	7891 → 49153 [PSH, ACK] Seq=1 Ack=1025 Win=8096 Len=1024
8	2.581868s	192.168.1.102	192.168.1.202	TCP	54	49153 → 7891 [FIN, ACK] Seq=1025 Ack=1025 Win=8096 Len=0
9	2.582253s	192.168.1.202	192.168.1.102	TCP	60	7891 → 49153 [ACK] Seq=1025 Ack=1026 Win=8095 Len=0
10	2.582361s	192.168.1.202	192.168.1.102	TCP	60	7891 → 49153 [FIN, ACK] Seq=1025 Ack=1026 Win=8095 Len=0
11	2.792432s	192.168.1.102	192.168.1.202	TCP	54	49153 → 7891 [ACK] Seq=1026 Ack=1026 Win=8095 Len=0

Fig. 5.15 TCP session with a data exchange of 1080 bytes with NFC channel sensing mechanism at 11.2 kbps and initial RTO of 5 s

No.	Time	Source	Destination	Protocol	Length	Info
1	0.000000s	MS-NLB-PhysServe…	Broadcast	ARP	42	Who has 192.168.1.202? Tell 192.168.1.102
2	0.000493s	MS-NLB-PhysServe…	MS-NLB-PhysServe…	ARP	60	192.168.1.202 is at 02:12:34:56:78:cd
3	0.000536s	192.168.1.102	192.168.1.202	TCP	58	49153 → 7891 [SYN] Seq=0 Win=8096 Len=0 MSS=1024
4	0.000853s	192.168.1.202	192.168.1.102	TCP	60	7891 → 49153 [SYN, ACK] Seq=0 Ack=1 Win=8096 Len=0 MSS=1024
5	0.079878s	192.168.1.102	192.168.1.202	TCP	54	49153 → 7891 [ACK] Seq=1 Ack=1 Win=8096 Len=0
6	0.603941s	192.168.1.102	192.168.1.202	TCP	1078	49153 → 7891 [PSH, ACK] Seq=1 Ack=1 Win=8096 Len=1024
7	0.604630s	192.168.1.202	192.168.1.102	TCP	1078	7891 → 49153 [PSH, ACK] Seq=1 Ack=1025 Win=8096 Len=1024
8	1.165876s	192.168.1.102	192.168.1.202	TCP	54	49153 → 7891 [FIN, ACK] Seq=1025 Ack=1025 Win=8096 Len=0
9	1.166360s	192.168.1.202	192.168.1.102	TCP	60	7891 → 49153 [ACK] Seq=1025 Ack=1026 Win=8095 Len=0
10	1.166529s	192.168.1.202	192.168.1.102	TCP	60	7891 → 49153 [FIN, ACK] Seq=1025 Ack=1026 Win=8095 Len=0
11	1.286891s	192.168.1.102	192.168.1.202	TCP	54	49153 → 7891 [ACK] Seq=1026 Ack=1026 Win=8095 Len=0

Fig. 5.16 TCP session with a data exchange of 1080 bytes with NFC channel sensing mechanism at 24 kbps and initial RTO of 3 s

sions and Dup ACKs which results in the reduction of latency. With a payload size of 1080 bytes, the TCP session latency is about 2.87 s at 11.2 kbps, which was 4.56 s before solving the retransmission problems. At 24 kbps the latency is around 1.33 s which was initially 2.45 s.

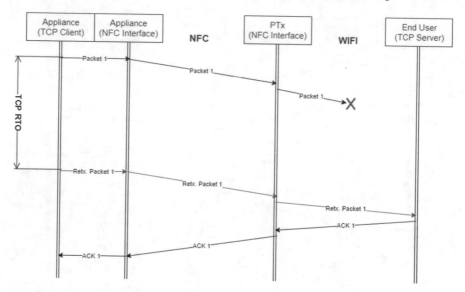

Fig. 5.17 TCP session with a very large TCP RTO

5.3.2   Avoiding TCP Spurious Retransmissions

5.3.2.1   Generalized Approach for TCP RTO Estimation

Setting a high initial TCP RTO will avoid spurious retransmissions for sure, however, it may also delay the retransmission when a packet is really lost. Figure 5.17 shows how the latency of the TCP session increases when a large initial TCP RTO is set and when packet loss occurs. It is therefore recommended to set the RTO slightly higher (at least one timer period) than the RTT of the data packet. This is because when the TCP stacks have coarse timers, there will be a tendency of timing out around one timer period sooner than what is estimated.

Now that all the retransmissions are removed, the RTT of the TCP packets can be estimated by analyzing the transmission conditions of the system in detail. The TCP/IP packet from the appliance travels through the NFC and Ethernet/Wi-Fi channels before reaching the end-user device. So the packet RTT can be broadly defined as

$$RTT = RTT_{NFC} + RTT_{WiFi} \text{ (ms)} \qquad (5.1)$$

where RTT is the total packet round trip time, RTT_{NFC} is the round trip time over the NFC channel and RTT_{WiFi} is the round trip time over the Wi-Fi channel.

The initial RTO set at compile time for standard wireless (or Ethernet) channels should be used as RTT_{WiFi}. Paxson et al. [6] recommend a minimum value of 1 s as

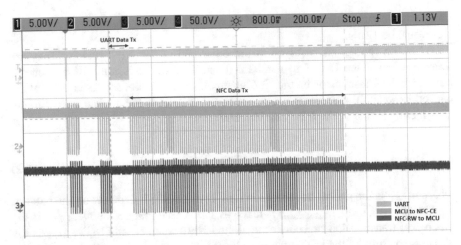

Fig. 5.18 TCP session capture in the direction from the appliance through the NFC-CE and NFC-RW modules at 11.2 kbps

the TCP RTO for wireless channels. The measured RTT of the previous packet can later be used to vary this value dynamically, as explained in the next section.

The RTT_{NFC} is the critical component which consumes most of the time. When the appliance stack transmits a packet, it first travels over the UART channel to reach the NFC module, as shown in Fig. 5.1. The NFC module then fragments the packet into chunks and transmits it over the NFC channel to the PTx stack. Figure 5.18 shows an oscilloscope output of a TCP session with 1080 bytes of data exchange at 11.2 kbps, captured between the ends of the NFC module. It depicts the packet transmissions in the direction from the appliance through the NFC-CE and NFC-RW modules. Different signals seen in the capture are explained below.

1. Yellow signal: it represents the transmission of data packets from the appliance stack to the MCU over UART.
2. Green signal: it represents the transmission of data chunks from the MCU to the NFC-CE module over UART.
3. Purple signal: it represents the transmission of data chunks from the NFC-RW module to the MCU over UART.

The time to transmit a fragmented 1080-byte data packet over the NFC channel is depicted as NFC Data Tx in Fig. 5.18. This value is equal to the theoretical time to transmit the data over the NFC channel, unless some time slots are missed in between. The UART Data Tx in the figure depicts the time needed to transfer the data packet from the appliance to the NFC module. This value adds to the packet processing time.

Apart from these, it is also important to take the waiting time for a time slot into account. From Sect. 5.1, it is clear that the packet chunks need to be present in the NFC module at least 2 ms before the arrival of the time slot. When the TCP/IP stack sends a packet, the packet can arrive at the NFC module at any point between two time slots. So the maximum amount of time a packet would need to wait for a time slot would be 12 ms.

When the chunks are received at the other end of the NFC channel, they are transmitted over the UART to the PTx stack. This transmission will be done in parallel to the transmission on the NFC channel, so they do not add to the RTT of the packet. However, the transmission of the last chunk needs to be taken into account. Considering all these, the RTT_{NFC} will be

$$RTT_{NFC} = 2 * (t_{UART} + t_{maxslotwait} + t_{NFC} + t_{UARTchunk}) \ (ms) \qquad (5.2)$$

where t_{UART} is the packet transmission time over UART. It depends on the baud of the UART being used. $t_{maxslotwait}$ is taken as 12 ms, as explained above. t_{NFC} is the theoretical transmission time over the slotted NFC channel, and $t_{UARTchunk}$ is the transmission time of the last chunk over UART.

The t_{UART} is given by the following equation:

$$t_{UART} = \frac{size_{pckt}}{baud_{UART}} \ (ms) \qquad (5.3)$$

where $size_{pckt}$ is the total packet size sent to the NFC module in bytes and $baud_{UART}$ is the baud of the UART in bytes per millisecond.

The t_{NFC} is given by the following equation:

$$t_{NFC} = slots_{pckt} * 10 \ (ms) \qquad (5.4)$$

$$slots_{pckt} = \frac{size_{pckt}}{size_{chunk}} \qquad (5.5)$$

where $slots_{pckt}$ is the number of time slots needed to transmit the packet. $size_{chunk}$ is the size of the payload section of the NFC protocol (in bytes). This depends or varies with the bit rate of the NFC being used.

The $t_{UARTchunk}$ is given by the following equation:

$$t_{UARTchunk} = \frac{size_{chunk}}{baud_{UART}} \ (ms) \qquad (5.6)$$

The initial RTO to be set must be greater than the maximum packet size that is transmitted through the NFC channel. In this experiment, the TCP MSS is set as 1024 bytes, which gives a maximum packet size of 1080 bytes. The total RTT for this packet size is estimated using Eq. 5.1, and it is found to be 3373.24 ms. As the timer period of the LwIP stack is 500 ms, this RTT value needs to be rounded up to the nearest 500 ms. This results in an optimum initial RTO of 3500 ms (3.5 s) for an NFC bit rate of 11.2 kbps. For the bit rate of 24 kbps, the optimum initial RTO is found to be 2.5 s.

5.3.2.2 New Algorithm for Dynamic TCP RTO Estimation

TCP in LwIP calculates the RTO after measuring the RTT of the data packets using Van Jacobson's (VJ) RTT estimation algorithm [7]. VJ's algorithm uses the Smoothed RTT (SRTT) calculation for RTO prediction. It measures the RTT value of the data packets to estimate the RTO of the next packet to be sent. Therefore, the RTO which is assigned to a packet is based on the RTT of the previous packet, which is done irrespective of the packet size.

Consider situations where the TCP sessions are long and the initial TCP RTO is set to 3.5 s at compile time (NFC bit rate of 11.2 kbps). For example, if a user chooses to cook step by step by creating their own recipe instead of uploading a recipe in one go, the TCP session would last very long and it could comprise several TCP messages with randomly varying sizes. If the application sends very small data packets of less than 10 bytes for a long time, TCP would adjust the RTO to a smaller value of about 1 s. Now, if the application suddenly sends very large packets, like recipes greater than 1 kB, an RTO of 1 s would be too small. This would result in spurious retransmissions until TCP adjusts the RTO according to the new packet sizes. On the contrary, if the application sends very small data packets right after sending large packets, the RTO of the small packets would be large initially until it is gradually adjusted to an appropriate value. In the meanwhile, if one of these packets gets lost, the system would take longer to timeout resulting in delayed retransmission (see Fig. 5.17). This would increase the overall latency of the system.

Figure 5.19 shows the TCP stream diagram of the client stack in a long session with 68 data packets of varying sizes. Points with the same sequence number denote retransmissions. It can be noticed that every time a large packet (denoted by a large jump in sequence number and/or time) is sent after a series of small packets, spurious retransmissions occur. This is because TCP would have adjusted the RTO suitable for small packets, and when large packets are suddenly sent this RTO would become too small considering the RTT of large packets. In the diagram, this is represented by packets having the same sequence number being sent more than once at different times. There are eight spurious retransmissions and eight Dup ACKs resulting in a total session duration of 22.58 s. It is very important to eliminate these retransmissions because it increases the latency of the TCP session in the order of seconds, due to the constrained nature of the NFC channel.

To mitigate this, a new algorithm is introduced that sets the TCP RTO depending on the estimated RTT of the current packet to be sent, instead of completely relying on the RTT estimation of the previous data packet. This approach has been designed by taking the following problems into account.

1. When the RTT estimation is made before sending the packet, the delay variability of the channels should also be considered. In VJ's RTT estimation algorithm, the RTO is adapted to the changing delay of the channel. If this mechanism is removed, then the stack will always assume a constant delay which may affect the latency by either causing spurious retransmissions or delaying retransmissions. Therefore, the new algorithm must take the delay variability into account.

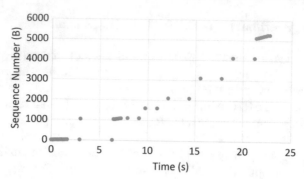

Fig. 5.19 Long TCP session with VJ's algorithm for setting the TCP RTO

Table 5.6 Problems due to delayed ACK and/or Nagle's algorithm

Packet size Sent/Received	Large	Small
Large	(Good) No spurious retx.	(Not bad) No spurious retx. but packet loss leads to delayed retx.
Small	(Bad) Spurious retx.	(Good) No spurious retx.

2. The Wi-Fi and the NFC channels could have variable delays. The NFC channel in the cordless kitchen would be used to send non-TCP/IP messages such as power control messages every now and then. This would affect the RTT of the TCP/IP messages and could delay their delivery. If the delay of the channel increases over time, it is difficult to identify if this increase is on the NFC channel or on the Wi-Fi channel. If the delay decreases from the theoretical value, it will be due to the reduced delay only on the Wi-Fi channel because the RTT on NFC will not go below the theoretical value (maximum reduction can be 10 ms, i.e. packet gets a time slot as soon as it arrives). So the RTO must be updated by closely observing the changing channel delay.

3. The RTT will be estimated considering the packet transmission time in both directions. The receiver may not always send back a packet of the same size. If only an ACK is received, the estimation will be larger than anticipated. But if the receiver replies with a bigger packet, for example, the delayed ACK algorithm does not send an ACK immediately, it waits for $<= 500$ ms [8] to check if the application has any further data to send so that it can piggyback the ACK with the next data packet. Another example is Nagle's algorithm which combines smaller packets to form a full-sized packet. In these cases, the estimated RTO will be smaller than the estimated value, which will lead to spurious retransmissions. Table 5.6 summarizes this problem. To solve this, the delayed ACK algorithm can be modified such that the stack sends an ACK for the received packet immediately, if the response packet is bigger than the received packet. This would avoid unnecessary spurious retransmissions.

4. If there is a packet loss, then that packet needs to be retransmitted using exponential backoff, where the RTO is doubled every time the same packet is retransmitted. For this, the estimated RTT of the packet needs to be used with the back-off multiplier.

Algorithm 1 New RTO estimation algorithm

1: RTT_p: Theoretical RTT of the previous packet
2: RTT_{meas_p}: Measured RTT of the previous packet
3: RTO_c: RTO of the current packet
4: RTT_{N_c}: RTT_{NFC} of the current packet
5: RTT_{W_c}: RTT_{WiFi} of the current packet
6: r: Factor r
7: expBackoff(): computes binary exponential backoff based on retransmit count
8:
9: **procedure**
10:     $r \leftarrow 1$ // Initialize r to 1
11:     **while** Packet queue is not empty **do**
12:         $RTT_{N\_c} \leftarrow$ Calculate theoretical RTT_{NFC} using Eq. 5.2
13:         $RTT_{W\_c} \leftarrow$ Use recommended initial RTO
14:         $RTT_p \leftarrow RTT_{N\_c} + RTT_{W\_c}$ // Store theoretical RTT to calculate r
15:         **if** $r >= 1$ **then**
16:             $RTT_{N\_c} \leftarrow r * RTT_{N\_c}$
17:             $RTT_{W\_c} \leftarrow r * RTT_{W\_c}$
18:         **else**
19:             $RTT_{W\_c} \leftarrow max(1000, r * RTT_{W\_c})$
20:         $RTO_c \leftarrow \lceil (RTT_{N\_c} + RTT_{W\_c})/500 \rceil * 500$ //Round-up to the next 500 ms
21:         **if** Retransmission = true **then**
22:             $RTO_c \leftarrow RTO_c *$ expBackoff() // Backoff procedure
23:         $RTT_{meas\_p} \leftarrow$ Measure and update RTT of the packet transmitted
24:         $r \leftarrow RTT_{meas\_p}/RTT_p$ // Compute r

Based on this analysis, the new algorithm is designed to dynamically estimate the optimum packet RTO value. The working of this algorithm is discussed in detail below.

- The theoretical RTO is calculated for each packet before its transmission, using Eq. 5.1. RTT_{WiFi} is set according to the initial RTO recommended for Wi-Fi (or Ethernet) channels. Furthermore, a minimum RTO value of 1 s is maintained for RTT_{WiFi} as recommended by [7]. The RTT_{NFC} is calculated as explained previously, using Eq. 5.2.
  (Note: The LwIP stack uses an initial RTO of 3 s. However, as the experiments are carried out on an Ethernet channel with <1 ms delay, an initial RTO of 1 s is used in the experiments.)
- The RTT of each data packet transmitted is dynamically measured to estimate the current delay in the NFC and Wi-Fi (or Ethernet) channels. The delay is estimated by comparing the theoretical RTT of the previous packet with the measured RTT

of the previous packet. The factor (r) by which the measured value varies from the theoretical value is calculated.

$$r = \frac{RTT_{measuredPrev}}{RTT_{Prev}} \tag{5.7}$$

where r is the ratio of measured RTT to theoretical RTT of a packet, $RTT_{measuredPrev}$ is measured RTT of the previous packet and RTT_{Prev} is the RTT of the previous packet calculated using Eq. 5.1.

- When the factor $r \geq 1$, the theoretical values of both RTT_{NFC} and RTT_{WiFi} are scaled up by this value. If r < 1, then only RTT_{WiFi} is scaled down. As explained earlier, this is due to the fact that the RTT of a packet over the NFC channel cannot go lower than its theoretical value. A minimum value of 1 s is maintained for RTT_{WiFi} as recommended by [7]. The new RTT is calculated with these scaled values using Eq. 5.1. For better estimation of the delay, a window of recent values of r can be maintained and the highest value in the window can be used for the current RTT estimation. The window size should be chosen depending on the type of applications being supported and the rate of packet transmission.

$$RTT_{NFC} := r * RTT_{NFC} \text{ (ms) } if \ r \geq 1 \tag{5.8}$$

$$RTT_{WiFi} := max(1000, r * RTT_{WiFi}) \text{ (ms)} \forall \ r \tag{5.9}$$

- The LwIP stack has a timer period of 500 ms to check for retransmission timeout. The RTO is therefore calculated as a multiple of 500 ms. So in the new RTO estimation algorithm, the estimated RTO of the current packet is rounded up to the nearest 500 ms.
- In case of packet loss, the exponential backoff algorithm is used with the estimated RTO of the lost packet. Using the estimated RTO for the backoff procedure would be more accurate than using the RTO of the most recently sent packet.
- To solve the spurious retransmission problem described in Table 5.6, the delayed ACK algorithm is modified such that an empty ACK will be sent if the size of the received packet is less than the size of the packet to be transmitted. A drawback of this solution is that the stack would send ACK packets even if the received packet size is slightly smaller than the packet to be sent. The RTO values are rounded up to the nearest 500 ms, so the packets of similar sizes may (but not necessarily) have the same RTO value. In this case, it would be unnecessary to send an extra ACK packet which could increase the latency of the system. It would be safe to use the modified algorithm even though it may not give the best result in the case discussed above.

Algorithm 5.1 summarizes the procedure for RTO estimation. It is tested on the TCP session shown in Fig. 5.19. The same experimental setup with the Ethernet channel is used for testing. Without the modification in the delayed ACK algorithm, a latency of 15.96 s is achieved as shown in Fig. 5.20, which is 6.62 s less than that with

Fig. 5.20 Long TCP session with new algorithm for RTO estimation

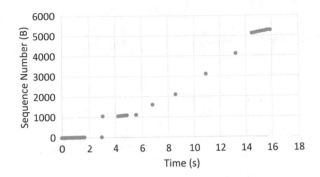

the original algorithm. This gives a 29.32% reduction in the latency in this example. However, there is still one spurious retransmission and one Dup ACK caused due to the delayed ACK algorithm. When the modified delayed ACK algorithm is used, all of the retransmissions are removed but the overall latency will be 16.1 s, which is slightly higher than the previous case. This is due to the fact that the stack sends out an ACK even when the received packet is sightly smaller than the packet to be sent. Note that the percentage improvement in the case of the new RTO estimation algorithm solely depends on the data set that is in consideration. It varies with different data sets.

References

1. lwIP - A Lightweight TCP/IP stack - Summary [Savannah]. (n.d.). Savannah. https://savannah. nongnu.org/projects/lwip/. Accessed 6 June 2021
2. *TCP/IP Illustrated* (3 Volume Set) by W.R. Stevens, G.R. Wright (2001) Hardcover. (2021). Addison-Wesley Professional
3. T. Klein, K. Leung, R. Parkinson, L.G. Samuel, Avoiding spurious TCP timeouts in wireless networks by delay injection. IEEE Global Telecommunications Conference, 2004. GLOBECOM '04., 5, 2754-2759 vol.5 (2004)
4. G.I. Fotiadis, V. Siris, Improving TCP throughput in 802.11 WLANs with high delay variability, in *2005 2nd International Symposium on Wireless Communication Systems* (2005), pp. 555–559
5. K. Leung, T. Klein, C. Mooney, M. Haner, Methods to improve TCP throughput in wireless networks with high delay variability [3G network example], in *IEEE 60th Vehicular Technology Conference, 2004.* VTC2004-Fall, vol. 4 (2004), pp. 3015–3019
6. V. Paxson, M. Allman, H.K. Chu, M. Sargent, Computing TCP's Retransmission Timer. RFC **6298**, 1–11 (2011)
7. V. Jacobson, Congestion avoidance and control. SIGCOMM (1988)
8. M. Allman, V. Paxson, E. Blanton, TCP Congestion Control. RFC **5681**, 1–18 (2009)

Chapter 6
Evaluation of the Bridge Architecture

Chapter 5 gave an overview of some of the factors influencing the performance of the bridge architecture and discussed how the standard TCP/IP stack can be adapted to the time-slotted NFC channel. Two major problems related to packet drops and spurious retransmissions were identified as the major contributors to the system latency. They were solved by introducing an NFC channel sensing mechanism and a new way of estimating and updating the TCP RTO values. This chapter contains the verification results of these solutions. Additionally, this chapter provides some recommendations for implementing the bridge architecture.

6.1 Implementation Recommendations

Certain use-case scenarios that are encountered while implementing the bridge architecture are listed below. Possible methods to handle/implement such scenarios are also briefly described. This research, however, does not consider these scenarios while evaluating the performance of the bridge architecture.

1. **Non-identical NFC buffer and MTU sizes in PTx and appliance:**
   The appliances and the PTxs may have different versions of software implementations, and they could be from different manufacturers. So it is not necessary that the uplink and downlink characteristics of the communication channel between the two will be the same.
   The PTx and appliance may have different buffer sizes in their NFC modules. Before starting a TCP connection, it is necessary to exchange information regarding buffer sizes so that packets with appropriate sizes can be sent without causing buffer overflows. It is also important for the appliance to know the Maximum

Transmission Unit (MTU) size of the PTx. The TCP MSS size can then be adjusted accordingly to prevent packet drops.

2. **Increased communication overhead due to small packet buffer size:**
The memory allocated for the TCP/IP packets by the stack should be large enough to hold an entire packet with maximum segment size. In the LwIP stack, a single TCP/IP packet is stored in multiple small packet buffers that are chained together. This type of storing increases the overhead in the packet and adds to the latency on the NFC channel.

3. **Upgrading the TCP/IP stack in the end-user devices:**
The new algorithm proposed for handling the RTO mechanism requires modifications to be made in the TCP/IP stack. This would be easy for the appliance because its stack needs to support only the NFC-enabled kitchen applications. On the contrary, the end-user device cannot readily make these changes as its TCP/IP stack is shared by various other applications. The stack needs to be upgraded with the new algorithm such that it dynamically supports all types of applications and channels.
As explained in Sect. 5.3.2.2, the algorithm sets the RTO of the packets by considering the NFC transmission rate, packet size and observing the delay on the Ethernet/Wi-Fi channel. Similarly, this method could also be used for applications that do not involve NFC channels. The stack can study the channel delay by constantly measuring the RTT of the packets and use this to calculate the delay experienced per byte on the channel. It can then set the RTO of the packets using the current packet size and the delay per byte factor. A better RTO estimation can be achieved with this method which would help in avoiding spurious and delayed retransmissions especially in high delay, low bandwidth channels. By upgrading the TCP/IP stack with this algorithm, it can dynamically adapt itself to different channels and support a wide variety of applications with improved performance.

6.2   Results

The performance of the bridge architecture is evaluated by carrying out various experiments with different NFC bit rates and data sizes. The performance is analyzed by measuring latency, throughput, number of retransmissions in the TCP sessions, NFC channel bandwidth utilization, etc.

6.2.1   Packet Retransmissions

Tables 6.1 and 6.2 show the number of retransmissions, DUP ACKs and keep-alive messages in the TCP session after using the mitigation techniques Sects. 5.3.1 and 5.3.2 described in Chap. 5, at 11.2 kbps and 24 kbps, respectively. The retransmitted

Table 6.1 Number of retransmissions at 11.2 kbps

NFC payload size (Bytes)	Retransmissions in TCP session			
	Original TCP/IP configuration	NFC channel sense	Optimum TCP RTO	NFC channel sense + Optimum TCP RTO
250	1R	1R + 1DA	1R	0R
500	2R	2R + 1DA + 2KA	1R	0R
1000	3R + 1DA	3R + 2DA + 2KA	1R	0R
1080	3R + 1DA	3R + 2DA + 2KA	1R	0R

Table 6.2 Number of retransmissions at 24 kbps

NFC payload size (Bytes)	Retransmissions in TCP session			
	Original TCP/IP configuration	NFC Channel sense	Optimum TCP RTO	NFC Channel sense + Optimum TCP RTO
250	1R	0R	1R	0R
500	1R + 1DA	0R	1R	0R
1000	1R	2R + 1DA + 2KA	1R	0R
1080	3R + 1DA	2R + 1DA + 2KA	1R	0R

packets are depicted by the symbol 'R', DUP ACKs by 'DA' and keep-alive packets by 'KA'. The experiments are carried out with TCP sessions exchanging single packets with NFC payload sizes of 250 bytes, 500 bytes, 1000 bytes and 1080 bytes at 11.2 kbps and 24 kbps.

The technique Sect. 5.3.2 introduced is to remove the spurious retransmissions by setting optimum initial RTO values for all the outgoing TCP/IP packets. Tables 6.1 and 6.2 show that by using only this solution, the total number of retransmissions can be brought down to one. The technique in Sect. 5.3.1 is an NFC channel sensing mechanism introduced to avoid packet drops at the NFC interface. As shown in the tables, using only technique in Sect. 5.3.1 the total number of packets increases compared to the respective original TCP sessions in most of the cases. However, when both these techniques are used together, all types of retransmissions, DUP ACKs and keep-alive packets are removed. Before concluding on the performance based on the number of packets in the TCP session, it is important to study the latency of the session, which is done in Sect. 6.2.2.

Figure 6.1 depicts the RTO values estimated by the new algorithm (Sect. 5.3.2.2) in a long TCP session with randomly varying packet sizes. These values are compared with the ones estimated by VJ's algorithm used in the LwIP stack and the packet RTT values obtained over an Ethernet channel with <1 ms delay. The estimations are, however, still carried out considering the Wi-Fi channel characteristics with a

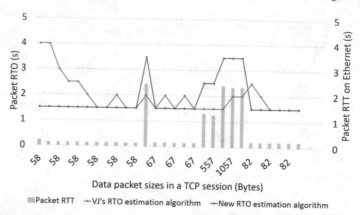

Fig. 6.1 Comparison of packet RTO values with the new and VJ's RTO estimation algorithms

minimum RTO of 1 s, which results in an offset of about 1 s between the measured RTT and the estimated RTO values as seen in Fig. 6.1. The new algorithm clearly gives a more accurate estimation of the RTO values compared to VJ's algorithm as it takes the packet sizes and bit rates of the channels into account, therefore avoiding all the spurious and delayed retransmission scenarios.

6.2.2   Latency

Reduction in the number of packets in the TCP session need not necessarily improve the latency of the session. This is because the time delay between packet generation, especially in the case of retransmitted packets, is also an important factor that affects the overall latency. Figures 6.2 and 6.3 show the graphs of latencies of TCP sessions with and without the mitigation techniques in Sects. 5.3.1 and 5.3.2 at 11.2 kbps and 24 kbps, respectively.

The percentage by which the latencies increase or decrease using the mitigation techniques compared to the original latency is indicated in the graphs. At lower NFC bit rates, for example, 11.2 kbps, the TCP session latency with only technique in Sect. 5.3.1 becomes higher than that with only technique in Sect. 5.3.2 when there are more number of retransmissions/DUP ACKs/keep-alive messages. This is because even though the packet drops are prevented, there will be too many extra packets to be transmitted over a low bandwidth channel. On the contrary, at higher bit rates like 24 kbps, the latency with only technique Sect. 5.3.1 will be lower than that with only technique Sect. 5.3.2 because when only technique in Sect. 5.3.2 is used, the time delay created by retransmission caused due to packet drop will be more significant compared to the packet transmission time on a relatively higher bandwidth channel. The TCP/IP stack has to wait for the timeout to realize that the packet is dropped and

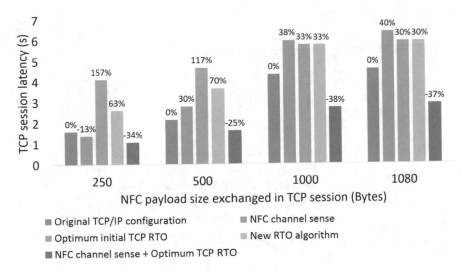

Fig. 6.2 Latencies of TCP sessions at 11.2 kbps

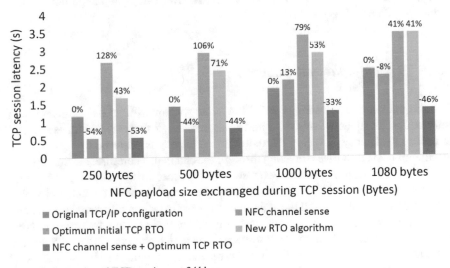

Fig. 6.3 Latencies of TCP sessions at 24 kbps

resend it. This waiting time will be longer compared to the time taken to transmit extra packets.

To achieve the best results, it is recommended to use both the mitigation techniques together. Using both, up to 38% reduction in latency can be achieved at 11.2 kbps and up to 53% at 24 kbps. Higher reduction is achieved at higher bit rates because of the same reason explained above. At higher bit rates, the time delay created because of packet drops will be more significant when compared to the total transmission

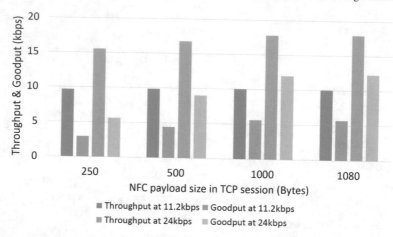

Fig. 6.4 System throughput at 11.2 and 24 kbps

time. So by removing this delay which is a bigger overhead, a higher gain in latency reduction can be achieved.

6.2.3 Throughput and Goodput

The throughput of the system remains the same with or without the retransmission mitigation techniques in Sects. 5.3.1 and 5.3.2. It is known that the techniques are used to reduce the latency, however, the reduction in latency is achieved by reducing the number of packets or bytes traveling through the channel. Therefore, the throughput, which is the number of bytes transferred per unit time, will be unchanged because with the mitigation techniques less packets/bytes travel through the channel which takes less time. So the overall throughput of the system technically remains constant.

Figure 6.4 depicts the throughput versus goodput of the system for different NFC payload sizes exchanged in the TCP session using both techniques in Sects. 5.3.1 and 5.3.2 at 11.2 kbps and 24 kbps. On an average, the throughput is 9.9 kbps at an NFC bit rate of 11.2 kbps and 17.01 kbps at 24 kbps. It can be seen that the goodput of the system improves with an increase in the payload size. This is because the overheads from the TCP/IP header and UI protocol become less significant with an increase in payload size. Choosing a bigger TCP MSS size will help in increasing the goodput of the system.

The throughput is lower for TCP sessions with small payload sizes, and it gradually increases with the increase in payload size. This is because with small payload sizes, the time spent waiting for a time slot will be significant compared to the packet transmission time. At higher bit rates, this becomes more noticeable because the transmission time will be even smaller. This affects the total transmission time and thus the throughput of the TCP session. The throughput could be improved by

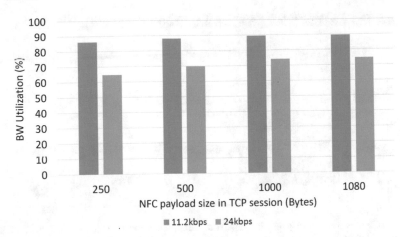

Fig. 6.5 Bandwidth Utilization at 11.2 and 24 kbps

1. using TCP/IP header compression techniques such as [1, 2].
2. employing the 6LoWPAN technology for the compression of TCP/IP packets over NFC as described in [3, 4].
3. letting the PTx detect and filter out the spuriously retransmitted packets and DUP ACKs coming from the appliance and the end-user device, similar to the technique proposed in [5]. This would reduce the number of packets on the NFC channel and improve the system performance.
4. modifying the NFC protocol in order to optimize the NFC handshake sequence as suggested in [6].

6.2.4 Bandwidth Utilization

The bandwidth utilization of the NFC channel for the experiments performed is illustrated in Fig. 6.5. It is calculated using the following equation:

$$BW\ Utilization = \frac{Throughput}{Theoretical\ BW} * 100\ (\%) \qquad (6.1)$$

The average bandwidth utilization is found to be 88.4% at 11.2 kbps and 70.89% at 24 kbps. Lower bandwidth utilization is observed at higher bit rate because the processing time which includes packet transmission time over the UART and Wi-Fi/Ethernet channels, packet processing time by the stack, etc. remains constant irrespective of the NFC bit rate. This processing time overhead will be more significant at higher bit rates because it has smaller NFC transmission time. When Figs. 5.18 and 6.6 are compared, it can be seen that the UART Data Tx time is the same at both 11.2 kbps and 24 kbps, which is a fixed overhead. However, NFC Data Tx time is

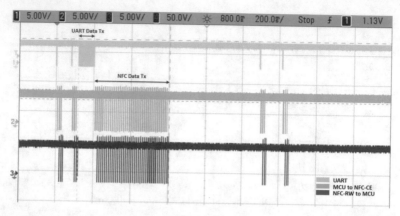

Fig. 6.6 TCP session capture in the direction from the appliance through the NFC-CE and NFC-RW modules at 24 kbps

smaller at 24 kbps compared to 11.2 kbps. This keeps the NFC channel idle for a longer period at a higher bit rate, thus reducing the bandwidth utilization. The main factors affecting the NFC bandwidth utilization of this system are

1. Packet processing time: The NFC channel remains idle while the TCP/IP packet is being processed and transferred over the UART from the stack to the NFC module.
2. Synchronization of data transfer with the communication time slot: The packet arrival time at the MCU can lie anywhere between two consecutive time slots. As explained earlier, the NFC module requires the packet to be available for transmission at least 2 ms before the time slot occurs. This may result in a waiting time of up to 12 ms for every packet (assuming that the subsequent chunks arrive on time), which adds to the total packet transmission time.

Some ways to improve bandwidth utilization are listed below. These techniques could not be tested due to the limitations in the available hardware.

1. Parallelizing the packet processing and packet transmission operations;
2. Increasing bit rate of serial communication (UART);
3. Eliminating the MCUs and directly interfacing the appliance and the PTx stacks to their respective NFC devices. This will reduce the processing delay caused by the serial communication.

References

1. V. Jacobson, Compressing TCP/IP headers for low-speed serial links. RFC **1144**, 1–49 (1990)
2. M. Degermark, M. Engan, B. Nordgren, S. Pink, Low-loss TCP/IP header compression for wireless networks. MobiCom '96 (1996)

3. J. Youn, Y. Hong, D. Kim, J. Choi, Y. Choi, Transmission of IPv6 packets over near field communication (2000)
4. J. Park, S. Lee, S. Bouk, D. Kim, Y. Hong, 6LoWPAN adaptation protocol for IPv6 packet transmission over NFC device, in *Seventh International Conference on Ubiquitous and Future Networks* (2015), pp. 541–543
5. Y. Kim, D. Cho, Considering spurious timeout in proxy for improving TCP performance in wireless networks. Comput. Netw. **44**, 599–616 (2004)
6. H. Sakai, A. Arutaki, Protocol enhancement for near field communication (NFC): future direction and cross-layer approach, in *Third International Conference on Intelligent Networking and Collaborative Systems* (2011), pp. 605–610

Chapter 7
Parametric Analysis of the Bridge Architecture

Chapter 5 discussed how the TCP RTT and RTO parameters can be adapted to the cordless kitchen system. The results after the adaptation were shown in Chap. 6. This chapter focuses on analyzing the TCP MSS and CWND parameters, and also other factors that affect the latency of the system. Simulations and theoretical calculations have been made to analyze the effects of parameters like the NFC bit error rate, communication time-slot size, frequency of non-TCP/IP messages over the NFC channel, etc.

7.1 Effect of TCP CWND Size and Slow Start Process on the System Latency

The TCP congestion control mechanism controls the maximum amount of data a sender can transmit before receiving an ACK from the receiver. The sender maintains a Contention Window (CWND) to keep track of this. The TCP congestion control consists of the slow start and congestion avoidance mechanisms, as shown in Fig. 7.1. The TCP slow start mechanism starts with an initial minimum CWND size and increases the window size by 1 MSS for every ACK received, until the slow start threshold (ssthresh) is reached. The congestion avoidance mechanism then takes over and gradually increases the window size until the network's capacity is reached or until a packet loss occurs. If it encounters a packet loss, the slow start process starts over with the minimum CWND size and with ssthresh set to half of the current CWND, as shown in Fig. 7.1 (Note: Refer to [1] for a detailed explanation on the working of the TCP/IP protocol).

It can be hypothesized that if the initial CWND is very small, then the latency of the TCP session would increase as the slow start process would take longer to

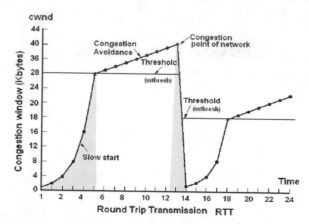

Fig. 7.1 TCP slow start and congestion avoidance mechanisms (*Source* [2])

Table 7.1 LwIP configuration for the TCP CWND experiments

TCP MSS	1024 bytes
Initial CWND size	4096 bytes
Maximum CWND size	8192 bytes
Sender buffer size	8192 bytes
NFC bit rate in slotted mode	11.2 kbps

reach the maximum window limit, making the channel idle for a significant amount of time. This hypothesis is tested with the LwIP configuration given in Table 7.1, and with a data transfer of 50 kB from the end-user device to the appliance. In the slow start process, for every ACK received, the CWND increases by 1 MSS. In case of packet loss, the ssthresh is set to half of the current CWND and the slow start process begins with initial CWND size of 1 MSS. This experiment is performed to find the optimum initial CWND size for the system such that there is minimum latency considering the congestion on the Ethernet/Wi-Fi channels.

Figure 7.2 shows the result of transferring 50 kB of data for different initial CWND sizes of 2048 bytes, 4096 bytes and 8192 bytes. It can be noticed that the difference in latency between the two extreme sizes is only about 76.07 ms. This implies that choosing a higher initial CWND size will not give a significant improvement in performance. The reasons for this behavior are explained below.

1. The NFC channel is half-duplex and has a low bandwidth in the time-slotted mode. So the bandwidth utilization of the NFC channel will be already high considering the small delay on the Ethernet channel and the high speed of the TCP/IP stacks. The data packets are almost always available to the NFC module unless the initial CWND is less than 2*MSS. Therefore, the reduction in the latency obtained by opting for a high initial CWND size will be very insignificant, as the bandwidth of the NFC channel cannot be improved further by a large factor.

Fig. 7.2 TCP session latency for different initial CWND sizes

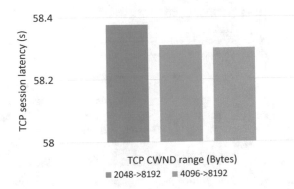

2. If the delay on the Ethernet/Wi-Fi channel is higher than or comparable to that of the NFC, the effects of a small initial CWND can be noticed. This is because the NFC channel may sometimes be idle when the packet is slowly traveling over the Ethernet/Wi-Fi channel. In this case, if it is made sure that there is at least one packet available at the NFC module at any point in time, it is possible to achieve maximum bandwidth utilization. So it is not necessary to always go for the maximum initial CWND size.

The experiment is repeated by varying the delays on the Ethernet channel to check if smaller initial CWND increases the latency by a significant amount. Figure 7.3 shows the results for Ethernet delays of <1 ms, 250 ms, 500 ms and 1 s, for different initial CWND sizes. It can be noticed that as the delay on the Ethernet increases, the difference in the latencies between the maximum and minimum initial CWND sizes increases. At an Ethernet delay of 1 s, there is a 1.08 s difference in the overall latency. This is not a very high gain in the performance though. Furthermore, the latencies with initial CWND sizes of 4096 bytes and 8192 bytes are almost the same. This implies that the NFC bandwidth utilization reaches the maximum with the initial CWND size of 4096 bytes. Any further increase will not result in any improvement.

Figure 7.4 shows the goodput graphs of the TCP sessions for different initial CWND sizes for a data transfer of 50 kB. It is interesting to see that no matter what the initial CWND size is, the goodput eventually comes to be 1 kbps or 8 kbps, which is the maximum achievable goodput on the NFC channel at 11.2 kbps, considering the NFC chunk size of 14 bytes with 10 bytes of usable payload size. This result supports the fact that the size of the TCP CWND does not have much effect on the throughput of the cordless kitchen system.

The TCP slow start process takes place only at the beginning of the TCP session if no packet loss is observed. So as long as there are no retransmissions, the effect of a small CWND may not be noticed in the system. This may not be true when retransmissions are taken into account, because every retransmission triggers the slow start process, making the TCP start over with a small initial CWND size. This could affect the overall latency. To verify this hypothesis, an experiment is designed where the end-user transfers 100 kB of data to the appliance, and the channel is lossy, where one out of 25 packets is lost. Ethernet delay of <1 ms is considered

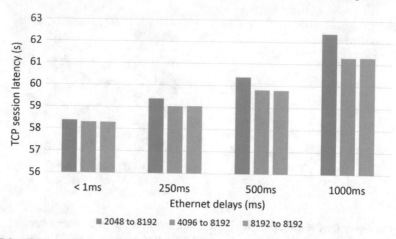

Fig. 7.3 TCP session latency for different initial CWND sizes and Ethernet delays

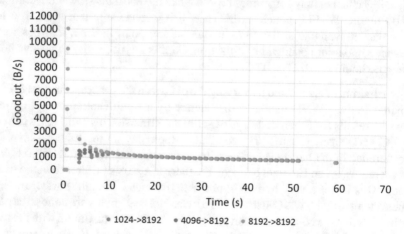

Fig. 7.4 Goodput of the TCP session for different initial CWND sizes for 50 kB data transfer

for this experiment. Figure 7.5 shows the latency for different initial CWND sizes. A reduction of up to 7.5% can be achieved when a bigger CWND size is chosen. Therefore, it can be concluded that the size of the initial CWND does not have a significant effect on the latency of the cordless kitchen system. This is because the NFC channel has a very small bandwidth and has almost maximum utilization even with small window sizes. So larger CWND does not help in further increasing the utilization of the channel. It should be noted that the intention of the TCP slow start process is to avoid congestion in the channel. It is recommended not to choose very high CWND as it may aggravate the latency in lossy congested channels.

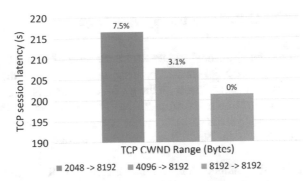

Fig. 7.5 TCP session latency for different initial CWND sizes over a lossy channel

7.2   Effect of TCP MSS Size on the System Latency

The TCP segments carry the actual data that is being transmitted. Choosing the right maximum segment size is very important to achieve minimum latency. If the MSS is very large, the size of the IP datagram will increase which may cause IP fragmentation reducing the efficiency of transmission. It may also increase the chance of the TCP segment getting lost. If the MSS size is too small, it would create more number of packets with very small data in each. In this case, the TCP/IP header overhead would become very prominent resulting in inefficient use of the channel bandwidth, thus increasing the latency.

Table 7.2 summarizes the average results of transferring 5 kB of data from the end-user device to the appliance using different TCP MSS sizes at 11.2 kbps. The results show that unless a very small MSS (<512 bytes) is chosen, the latency will not increase by a large number. A small MSS of 256 bytes increases the latency by 26.28%, however, choosing a size greater than or equal to 512 bytes increases the latency only by <10%. So an MSS value of 1024 bytes or greater would give a very high performance with minimal latency. It is important to note that in the case of an erroneous NFC/Wi-Fi channel with a high bit error rate (BER), large packets would be more susceptible to errors compared to smaller packets. So the TCP MSS should be chosen depending on the conditions of the channel in order to avoid retransmissions caused by packet errors.

Table 7.2   TCP session latency for different TCP MSS values

TCP MSS (Bytes)	1460	1024	512	256
TCP session latency (s)	6.24	6.29	6.78	7.88

7.3   Effect of NFC BER on the System Latency

The bit errors in the NFC channel would introduce errors in the TCP/IP packets being tunneled through the NFC channel causing the packets to be dropped by the TCP/IP stacks due to failing checksum. In the given setup, there is no error detection or correction mechanisms implemented in the NFC layer. So even a single bit error in the packet would lead to retransmission as the packet will be dropped. Therefore, the presence of bit errors in the NFC channel will have a huge impact on the system latency.

Bit errors in the NFC channel can be random or bursty. Random errors would lead to more number of packet drops as the errors are randomly distributed, which can affect any packet in the TCP session. On the contrary, the burst errors come as a block, so the errors would be confined to a single or a couple of packets depending on the size of the block and the time of occurrence. So the burst error would have less impact on the TCP session latency compared to random error. An experiment is designed to verify this hypothesis where the appliance and end-user device exchange 100 packets of 500 bytes each. Random and burst errors of 10^{-4}, 10^{-5} and 10^{-6} are introduced in the NFC channel to test the latency of the TCP session.

7.3.1   Random Errors

The random bit errors are modeled using the following formula:

$$P(0) = 1 - P(1) \tag{7.1}$$

where P(0) and P(1) denote the probabilities of transmitting bits 0 and 1 without errors.

7.3.2   Burst Errors

The burst errors are introduced based on the Gilbert–Elliott model as shown in Fig. 7.6. The states Good and Bad represent the bit error conditions. p and q are the transition probabilities between these states. The average burst length is taken as 4 bits in this experiment. An example below shows the steady state and transition probabilities of the Good and Bad states for an NFC BER of 10^{-4}.

$$\pi_{Good} = \frac{q}{q + p} \tag{7.2}$$

$$\pi_{Bad} = \frac{p}{p + q} \tag{7.3}$$

Fig. 7.6 Gilbert–Elliot error model for simulating burst errors

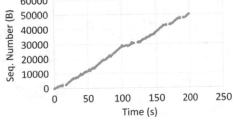

Fig. 7.7 TCP session with a random BER of 10^{-4}

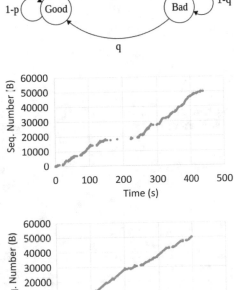

Fig. 7.8 TCP session with a burst BER of 10^{-4}

where $\pi_{Good} = 1 - 10^{-4}$ (steady state probability of state Good) and $\pi_{Bad} = 10^{-4}$ (steady state probability of state Bad). The transition probabilities p and q are calculated by taking Average Burst Length (ABL) as 4 bits. So q becomes 0.25 and p becomes $25 * 10^{-6}$.

Figures 7.7, 7.9 and 7.11 show the output of the TCP session with random NFC BERs of 10^{-4}, 10^{-5} and 10^{-6}, respectively, at 11.2 kbps. It can be seen that as the BER reduces, the number of retransmissions decreases and hence the TCP session latencies. The same behavior is observed with burst NFC BER as shown in Figs. 7.8, 7.10 and 7.12 at 11.2 kbps. As per the hypothesis, at a given NFC BER, fewer retransmissions are observed with burst errors compared to that with random errors. This proves that burst errors have less impact on the system latency as the errors come in bursts which affect fewer TCP/IP packets.

Table 7.3 summarizes the latencies of the TCP sessions with random and burst errors at different NFC BERs. At a BER of 10^{-4}, the latency with the burst error is around 54.56% less than that with random error. However, as the BER reduces, the difference in latency between the two types of errors reduces. At a BER of 10^{-6}, there is only about 1.8% difference in the session latencies. Therefore, it can be concluded that at lower BERs the type of error will not matter much but at higher BERs burst errors will have lesser impact on the overall latency.

Fig. 7.9 TCP session with a random BER of 10^{-5}

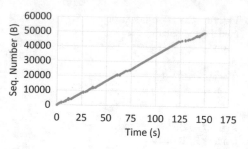

Fig. 7.10 TCP session with a burst BER of 10^{-5}

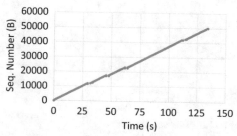

Fig. 7.11 TCP session with a random BER of 10^{-6}

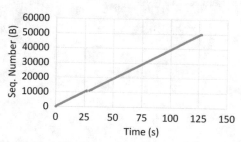

Fig. 7.12 TCP session with a burst BER of 10^{-6}

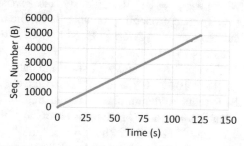

Table 7.3 TCP session latencies with random and burst errors at different NFC BERs

NFC BER	TCP session latency (s)	
	Random error	Burst error
10^{-4}	439.22	199.6
10^{-5}	150.07	134.46
10^{-6}	127.32	125.03

7.4 Effect of Varying the NFC Communication Time-Slot Duration on the System Latency

The cordless kitchen specification defines an NFC communication time-slot size of 1.5 ms in the time-slotted mode. If the size of the time slot is increased, the latency of the system can be reduced because more data can be transferred over bigger time slots. The decrease in latency with the increase in time-slot size will be non-linear because they are inversely proportional. An experiment is carried out to find out the optimum NFC time-slot size for the system such that minimum latency is maintained.

The standard payloads of NFC read and write commands, as defined in the cordless kitchen specification, are used in this experiment (refer to Sect. 2.5.2). A short TCP session exchanging 1 kB of data is considered, with NFC time-slot sizes varying from 1 to 2.5 ms. The latencies are theoretically calculated using RTT Eq. 5.1 as described in Sect. 5.3.2.1. Figures 7.13, 7.14 and 7.15 show the results of tunneling the TCP/IP packets over different time-slot sizes at 212 kbps, 424 kbps and 848 kbps, respectively. The graphs represent an inverse variation function. The rate at which the latency decreases with increasing slot size is steep at the beginning, and it gradually flattens out at higher slot sizes. This behavior is more noticeable at lower bit rates because the amount of data that can be sent over a time slot is small compared to that at higher bit rates. This implies that the choice of the correct time-slot size is more critical at lower bit rates for maintaining a reasonable latency.

In the cordless kitchen system, there is a trade-off between the efficiency of power transfer and communication. As the size of the time slot increases, the efficiency of the power transfer decreases. Moreover, bigger time-slot sizes would generate harmonics in the power signal leading to vibrations and heating in the PTx module. Therefore,

Fig. 7.13 Latencies of TCP sessions for different NFC time-slot sizes at 212 kbps

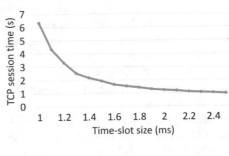

Fig. 7.14 Latencies of TCP sessions for different NFC time-slot sizes at 424 kbps

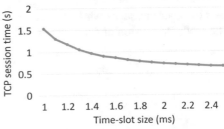

Fig. 7.15 Latencies of TCP
sessions for different NFC
time-slot sizes at 848 kbps

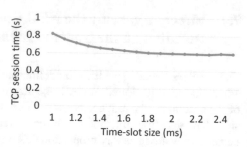

it is important to choose an optimum slot size such that both power transfer and data transfer are efficient. An optimum time-slot size needs to be chosen for the lowest NFC bit rate, which is 212 kbps in this case. This slot size would give a better performance at higher bit rates as well.

At 212 kbps, an efficiency of about 50% in data transmission can be achieved with a time-slot size of 1.5 ms. Higher efficiency with the same slot size can be achieved at higher bit rates. A slot size of 1.5 ms results in an efficiency of 75% at 424 kbps and about 95% at 848 kbps. An efficiency close to 99% can be achieved with a slot size of 1.9 ms at 848 kbps. Depending on the criticality of the Internet applications, appropriate slot size can be chosen such that desired efficiency is achieved at all data rates.

7.5   Considering Non-TCP/IP Messages over the NFC Channel

For ease of analysis, in all of the experiments the NFC channel was assumed to comprise only TCP/IP messages. However, in a real case scenario, the NFC channel would also carry other types of messages such as power control, negotiation, measurements and state transition. The frequency of these messages would depend on the type of application being used. An experiment is performed to analyze the performance of the TCP session in the presence of such messages at different frequencies of their occurrence.

The frequency of non-TCP/IP messages is taken as a fraction, for example, 2 slots every 10 slots used for other messages. This will impact the RTT of a TCP/IP packet over the NFC channel, i.e. the t_{NFC} in Eq. 5.4 will increase by the number of slots used by other messages while the packet is being transferred over the NFC channel. The modified version of Eq. 5.4 is as follows:

$$t_{NFC} = \frac{slots_{pckt}}{1 - freq_{ctrlmsgs}} * 10 \text{ (ms)} \tag{7.4}$$

Fig. 7.16 Latencies of TCP
sessions for different
frequencies of non-TCP/IP
messages at 848 kbps

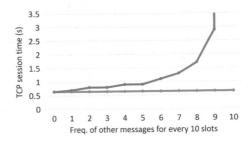

where

$$freq_{ctrlmsgs} = \frac{a}{b} \qquad\qquad (7.5)$$

a/b signifies 'a' slots every 'b' slots format. b is taken as section size and $b - a$ will be usable slots per section. The t_{NFC} has to be rounded up to the nearest section size b because the usable slots can occur anywhere in the section. As the frequency of other messages increases, the TCP session latency also increases. This increase will be non-linear because the number of usable slots varies inversely with latency. If $freq_{ctrlmsgs} = 0$, all slots will be available for TCP/IP packets and Eq. 7.4 will be the same as Eq. 5.4. If $freq_{ctrlmsgs} = 1$, i.e. all slots are used for other messages, then the t_{NFC} becomes ∞ which means that the TCP/IP messages cannot be transferred. A TCP session with 1 kB data exchange and a section size b of 10 slots is considered for the experiment. Theoretical calculations are made for TCP session latencies using Eq. 5.1 and for an NFC bit rate of 848 kbps.

The result of the experiment is depicted in Fig. 7.16. The graph shows an inverse variation function, so the rate of increase in latency will be steep as the frequency of non-TCP/IP messages increases. At lower frequencies, the latency varies slightly. It can be inferred from the results that efficiency of around 72% can be achieved in the transmission of TCP/IP messages at a frequency of 5/10. Appropriate frequencies can be chosen depending on the criticality of the Internet application.

References

1. *TCP/IP Illustrated* (3 Volume Set) by W.R. Stevens, G.R. Wright (2001) Hardcover (2021). Addison-Wesley Professional
2. G. Abed, M. Ismail, K. Jumari, A survey on performance of congestion control mechanisms for standard TCP versions. Aust. J. Basic Appl. Sci. **5**, 1345–1352 (2011)

Chapter 8
Conclusion

This research focused on enabling Internet connectivity to a new generation of smart kitchen appliances that work on wireless power technology. In order to provide efficient and seamless communication with the appliances, the possibilities of utilizing the NFC channel for Internet connectivity were explored. Two architectures called proxy and bridge were proposed to enable connectivity via a time-slotted NFC channel of the cordless kitchen. The bridge architecture was adopted for the cordless kitchen as it implements the full TCP/IP stack on the appliance and provides it more control over the TCP communication.

As most of the IoT applications use the TCP/IP protocol, this work mainly focused on adapting the TCP/IP protocol to the cordless kitchen system. Two major problems, namely spurious retransmissions and packet drops at the NFC interface, which arise while adapting TCP to the time-slotted NFC channel, were recognized and discussed in detail. To eliminate the spurious retransmissions, a generalized solution was provided to compute the optimum RTO values for TCP/IP packets tunneled over the NFC channel. Using this, the spurious retransmissions were completely removed in short TCP sessions. As TCP does not consider the payload sizes for RTO estimation, spurious retransmissions were observed when there was high variability in payload sizes. To mitigate this, a new algorithm was proposed that dynamically estimates and updates the RTO of the packets considering the payload sizes and changing channel delays. This algorithm provided a more accurate estimation of the RTO values compared to VJ's algorithm used in the LwIP stack. A reduction of about 29.32% in latency was observed with this algorithm for the data set considered. To avoid packet drops at the NFC module caused due to small inter-packet delays, an NFC channel sensing mechanism was introduced in the cordless kitchen system. This mechanism eliminated all the retransmissions that existed due to packet loss at the NFC interface. Using all these techniques, up to 38% reduction in the system latency is achieved at an NFC bit rate of 11.2 kbps and up to 53% at 24 kbps.

© The Author(s), under exclusive license to Springer Nature Switzerland AG 2021
S. Kashyap et al., *Cook Over IP*,
SpringerBriefs in Applied Sciences and Technology,
https://doi.org/10.1007/978-3-030-85836-0_8

   A comprehensive performance analysis was made with respect to different parameters of TCP/IP that affect the system performance such as the TCP RTO value, MSS
and CWND size. Take-aways of this study include (a) initial CWND size does not
have a significant impact on the system latency and (b) recommended MSS value is
≥ 1024 bytes to get a good performance. The research also analyzed the effects of
non-TCP/IP factors on the system performance such as NFC bit error rates, communication time-slot sizes and influence of transferring other control messages along
with TCP/IP packets over the NFC channel. It concluded that the system has relatively better performance with bursty errors in the NFC channel than with random
bit errors. It showed that close to 99% data transmission efficiency can be achieved
with a slot size of 1.9 ms at higher NFC bit rates. The results also showed that if
the frequency of non-TCP/IP messages on the channel stays under 50%, a transmission efficiency of 75% can be achieved. By adopting the proposed architectures with
improvement techniques and recommendations, Internet connectivity can be enabled
in the cordless kitchen system using its time-multiplexed NFC channel.

Reference

1. Gartner identifies top 10 wireless tech trends for 2019 and beyond (2019). Gartner. https://www.gartner.com/en/newsroom/press-releases/2019-07-23-gartner-identifies-the-
   top-10-wireless-technology-tre

Printed in the United States
by Baker & Taylor Publisher Services